# MEMORY

# MEMORY

EDITED BY

PHILIPPE TORTELL

MARK TURIN

MARGOT YOUNG

27 26 25 24 23 22 21 20 19 18 5 4 3 2 1

Printed in Canada on FSC-certified ancient-forest-free paper (100% post-consumer recycled) that is processed chlorine- and acid-free.

---

Cataloguing data available from Library and Archives Canada

ISBN 978-1-7752766-0-9 (softcover)
ISBN 978-1-7752766-1-6 (pdf)
ISBN 978-1-7752766-2-3 (epub)
ISBN 978-1-7752766-3-0 (Kindle)

Peter Wall Institute for Advanced Studies
University of British Columbia
Vancouver, BC
www.pwias.ubc.ca

Memory is the seamstress, and a capricious one at that. Memory runs her needle in and out, up and down, hither and thither.

– Virginia Woolf, *Orlando*

How can I begin anything new with all of yesterday in me?

– Leonard Cohen, *Beautiful Losers*

# CONTENTS

# MEMORY

# INTRODUCTION

*Philippe Tortell, Mark Turin, and Margot Young*

ON NOVEMBER 11, 1919, King George V of Britain inaugurated a tradition of remembrance for the fallen soldiers of the First World War. One year earlier, hostilities in the "Great War" had ceased, though the conflict would only officially end with the signing of the Treaty of Versailles in June 1919. In the early days after the war, and certainly during the first official ceremonies at Buckingham Palace, optimism and hope mixed with a profound sense of loss and grief. Surely, the world had learned an important lesson from this "war to end all wars."

As we know from the decades that followed, this was not to be. Rather, much of the world would soon find itself engulfed once again in a horrifying conflict that would turn out to be longer and bloodier than its predecessor, with particularly devastating impacts on millions of civilians. And conflicts continued over the twentieth century and into the current one. King George could scarcely have imagined the number of people who would come to be remembered on November 11.

Over the past century, the practice of Remembrance Day, as it is known throughout the Commonwealth, has changed significantly. There are no living survivors of the First World War, and the same will soon be true of the Second World War. Sadly, there is no shortage of those wounded and killed in war to take their place. What has changed, however, is the nature of global conflict and our understanding of the physical and psychological effects of war on combatants and civilians.

As we mark the one hundredth anniversary of the end of the First World War, we do so not only with diverse perspectives on what constitutes war but also with a sharpened sense of the injustices and trauma perpetrated in times of (and sometimes in the name of) peace. Take, for example, the appalling treatment of Indigenous communities in Canada at the hands of colonial settlers. This history is marked by the appropriation of lands, the destruction of entire communities through imported diseases, the disentitlement of and violence directed against Indigenous women, and the forced removal of thousands of children into residential schools, where they suffered physical and emotional abuse and the erasure of their cultural heritage. The Truth and Reconciliation Commission of Canada was established to shed light on these injustices, forging a collective consciousness of past wrongs and calling for institutional actions to create a fairer society. Meaningful reconciliation demands remembrance of the harms of colonialism, past and present.

As collective memory changes, so too does a society's ability to understand and express different facets of remembrance. Through legal and political systems and through the creative and performing arts, we have developed increasingly sophisticated and technologically mediated modes of recalling and revitalizing societal memory. With respect to individual memory, we know more than ever about the workings of our unconscious and about the complex neural networks that make up our brains. We are now beginning to understand the molecular mechanisms of how (and when) memories are formed. We are also advancing our understanding of nonhuman aspects of memory, whether through the creation of sophisticated digital technologies or through deep study of Earth history and the evolution of our universe, reaching back to the very beginning of time.

It was in this context that we took on the challenge of exploring memory from a wide range of perspectives. We sought to bring together a collection of voices from some of the world's leading thinkers and scholars in order to examine memory through a variety of lenses. The project began with a series of discussions in the fall of 2017, at the Peter Wall Institute for Advanced Studies at the University of British Columbia in Vancouver. The institute, which recently celebrated its first quarter century, brings innovative thinkers together to explore fundamental research questions from creative interdisciplinary perspectives. These initial discussions were as surprising as they were inspiring, and they helped us generate a list of topics that would eventually turn into the collection you now hold in your hands.

Our initial vision for this book, influenced by personal experience and professional training (in oceanography, anthropology, and law, respectively), was swept cleanly away. The contributions that we gathered address a range of topics far beyond our own disciplines, from molecular genetics and astrophysics to cultural history and the arts. We were challenged to seek unifying themes that illuminate and deepen our understanding of memory.

Important and transdisciplinary themes surfaced. Most significantly, perhaps, the essays share an appreciation of the fragility and fluidity of memory and its capacity to convey different meanings across time and space. The insights of neurological science into how memories are formed, archived, and retrieved help us better understand the tendency for human memory to change over time. Several authors discuss the malleability of memory and the implications of this for the administration of legal justice or for our understanding of shifting ecological baselines in the face of an expanding human footprint on natural systems. Some essays focus on the choices we make in commemorations and ceremonies that either reinforce or challenge dominant cultural narratives. In this light, it is critical to understand how memory has been transmitted over thousands of years through Indigenous oral histories, storytelling, and embodied cultural practices and to understand the role that museums and archives can play in shaping and reclaiming memories.

Long-term perspectives on memory give us a richer understanding of the world, challenging us to expand our imagination beyond our everyday experience. Several essays examine how the creative and performing arts can be used as a vehicle for transforming our understanding of past traumatic events. At the same time, the tools of modern science and technology have given us the capacity for seemingly limitless digital memory, while also creating a legacy of environmental destruction that lives on in the synthetic materials that have accumulated around the globe. Authors in this collection explore the memory of materials at the atomic level or, at much broader scales, how science enables us to read the memory of our planet, our place in the universe, and the spectacular events that led to the diversification of life on Earth.

In appreciating what it means to be human, understanding how, when, and why we forget is just as important as thinking about what we remember. Several of the contributions thus explore how memory can be harmful and what the loss of memory signals for our sense of self.

We hope this collection will challenge you to think creatively and deeply about memory – its composition and practice. We ask you to consider the meaning of memory for individuals and societies and for our material world. The essays in these pages offer a road map to explore the ways that memory matters and how it is transmitted, recorded, and shaped across space and time.

# HEALING THROUGH CULTURE

*Hilistis Pauline Waterfall*

THERE'S A HEILTSUK CHILD IN ME whose growth stopped at the age of twelve and doesn't know a Heiltsuk adulthood. Conversely, there's a white adult in me who started to grow at the age of twelve but doesn't know a white childhood. I left home at age twelve to attend school far from home and was thrust into a foreign world in complete cultural shock. Not having a good command of English compounded the problem. The Indian residential school system was imposed on us as First Nations children in Canada for at least 125 years. Its painful and negative results are documented through the Indian Residential Schools Truth and Reconciliation Commission, or TRC, which brought into broad public light the multigenerational cultural loss and displacement of at least 150,000 First Nations children, including me.

According to Murray Sinclair, the chair of the TRC, education caused this mess, and education will also get us out. As a survivor of this system, my indoctrinated mind equated education with Western values. I am now confident that the key to unravelling this mess lies in equating education with more traditional Heiltsuk values. Despite the intergenerational disruption in the transmission of knowledge, our collective cultural memory continues to exist, inform, and support the renewal and revitalization of Heiltsuk ceremonies, values, teachings, language, and beliefs. Decolonizing my programmed Western mind and embracing my Heiltsuk mind through traditional learning were key to reinforcing my sense of place, both within my Heiltsuk culture and the outer Western world.

Innate memory runs deep and, when nurtured and guided, helped me to heal and rekindle my confidence and cultural pride. The revival of cultural memory was fundamental in the steps taken to help rebuild our potlatch system over the past thirty years. Heiltsuk youth now participate freely in learning our Haíłzaqv language, in dancing, singing, drumming, and other cultural ways that were banned for at least six decades. Based on my experience, cultural renewal is key to continued adaptation and survival, and traditional education will continue to provide a pathway to healing and reconciliation.

My Heiltsuk childhood was warm, predictable, and safe. My parents nurtured and instilled Heiltsuk values, including the importance of honesty, respect, generosity, and sharing. The sense of family was strongly rooted. The spirit of community and belonging was solid. Immersion in Heiltsuk ways and language was my reality, as was a healthy diet of natural and abundant Heiltsuk traditional foods. Opportunities to learn and develop essential skills in age-appropriate tasks were provided. A sense of responsibility was instilled as was a sense of confidence and connectedness. Berry picking with my mother was important bonding time, giving me the satisfaction of contributing to my family's needs. Emerging conflict-resolution skills, social development, team work, and relationship building came from childhood and sibling relationships. All this was to change suddenly without adequate preparation or explanation.

My life was like a piece of cork tossed into the ocean at the mercy of storms and changing tides – alone, confused, afraid, disconnected, and lost. Navigating the outside world eclipsed my Heiltsuk roots and often left me caught between the cracks of these two worlds, not truly fitting into either. Formal education was fraught with preconceived racialized perceptions of my being academically handicapped. Despite pitfalls, I persisted and eventually attained my bachelor of education degree from the University of British Columbia at the age of forty-five. As a mature, and perhaps naive, student, I had assumed that the Western educational environment would be more welcoming and open. For the most part it was, but my fragile confidence still found it alienating and intimidating.

In the eighteen years during which I was removed from my Heiltsuk world, I had to adjust to a world that was the complete opposite of what I had known. As a mere number in residential school, I experienced confusion, fear, and isolation. Rather than the more rational *t'gai'la* (to give advice) that I had experienced as a Heiltsuk child, corporal punishment was a standard practice. *Responsibility* in such institutions meant menial chores such as scrubbing floors with a toothbrush as a consequence of trying to question harsh expectations that never seemed rational or reasonable. An inadequate and high-carbohydrate diet of unfamiliar and strange foods left me hungry and longing for the home-cooked, protein-rich diet that I was used to. Opportunities to further develop decision-making skills, family relationships, and healthy social awareness were almost nonexistent. The dormitories were filled with lonely, misplaced girls like me.

Between the time that I left the school and returned home, I travelled back and forth between Bella Bella and other places in British Columbia thirty-four times. I settled permanently in Bella Bella after eighteen years of being away. I was a stranger in my homeland, no longer fitting into my Heiltsuk world and with no sense of belonging to my family. Without any clues to guide my reintegration, and with a feeling of being "Heiltsuk ignorant," I began to ask questions, most of which must have seemed ridiculous to my parents and old people of the day. In retrospect, I now understand why my Heiltsuk teachers were so patient – they too had had alienating and difficult experiences. I later learned that a 98 percent dropout rate from schooling was an understandable response to an unhealthy situation.

Upon my permanent return to Bella Bella, I found my ninety-six-year-old great-grandmother entering the last stages of life. I intuitively felt that learning about her was a key to my Heiltsuk quest. As a tribute to her, I undertook a genealogy project to honour her. Not knowing that she had had twenty-two children, I became consumed with researching and documenting what I thought would be a simple family tree. More than half of her children had died when they were babies, but she had ten surviving adult children – all of whom were old-age pensioners with large families of their own. I discovered that after her husband had died, my great-grandmother became a resilient and independent widow who built three houses in her lifetime to accommodate her growing, extended family. She had 396 direct descendants spanning five generations, and I was her oldest great-grandchild. I documented her family tree in a booklet and distributed it at a feast to honour her memory. Imagine my surprise when other families asked me to help create their family trees. I was baffled until I realized just what a toll the removal of five generations of Heiltsuk children to residential schools had taken on family relationships across the community. Through memory and interviews over five years, we pieced together the intergenerational connections to nearly every coastal village, based on the previous practice of arranged marriages between offspring of hereditary chiefs. The process was both exhilarating and gratifying, as we learned together. What began as a personal quest to understand where I fit in ended up being an important healing outcome that continues today. These days, the host of every potlatch and feast documents maternal and paternal family trees, which are attached to the programs and distributed to guests. Consequently, there is now a strong understanding of who our *waa-waax-toos* (family relations) are, which is a key to continued healing of our fractured past.

In 1968, my non-Heiltsuk husband arrived in my community as a member of the Royal Canadian Mounted Police. He observed a striking lack of cultural identity or practices that defined us as Heiltsuk people and as a community. Because of the potlatch ban and the missionary hospital, church, and school, which had been in our village for a century, our Heiltsuk ceremonies and customs had nearly vanished. Nearly two decades later, in 1985, my husband observed how extended our mortuary processes were and asked if this was because we revered death. After my initial offence and shock in response to his question, I reached out to old people for an answer. I learned that during the potlatch ban and missionary influences, cultural practices went underground and were practised in private homes with windows blacked out during times of death. Apparently, the missionaries had compassion for our people during these times of mourning and allowed some mortuary practices to continue. To take advantage of this opportunity, nonmortuary practices were also conducted. To accommodate our "illicit" cultural practices in a nonpublic setting, some of our chiefs had built large two-storeyed homes: the main floor was made up of a big living room area that accommodated up to fifty or more guests, alongside a small kitchen to prepare food to feed guests. All bedrooms were built on the top floor. This was an ingenious solution to sustain some of our Heiltsuk practices.

In 1985, I asked my grandmother why customs such as coming-of-age practices were included during mortuary feasts. She was still under the impression that there was a law against potlatches, and I presume that there wasn't an official announcement when this policy was reversed. After our conversation, Granny and I began to talk about the various ceremonies and practices that would have taken place in a regular feast or potlatch setting and what took place in a mortuary feast. She taught me that *Nuyem* is a Heiltsuk word that describes the treasure chest of memories and knowledge that embodies and holds ancestral cultural inheritance through stories, dances, traditional names, history, and so on. We began to differentiate which practices were mortuary and which weren't and to reweave the threads of knowledge and ways based on memories of several knowledge keepers.

In that same year, a family member asked me to help put together a program for a potlatch – the first in one hundred years to be held in our community. Again, my ignorance of Heiltsuk ways required me to interview, question, and document the remnants of existing potlatch knowledge. I learned about ceremonies that included youth coming of age, mourning ceremonies, washing or purification ceremonies, traditional wedding ceremonies, and so on. Parallel to my cultural investigation, there was a Heiltsuk man who had returned home after being at residential school. He was an artist on a mission to learn Heiltsuk songs, dances, names, and so on. His mother was a seamstress who made button blankets and aprons with family animal crests. We began to put together the potlatch program based on the recollections of our knowledge keepers. The potlatch was a success and sparked a cultural flame.

Ironically, it was work by the German American anthropologist Franz Boas at the turn of the twentieth century that created a valuable resource used to trigger memories of Heiltsuk. Although incomplete, the stories he gathered helped to rebuild the memory bank needed to piece together a semblance of the potlatch held in 1985. This fragile beginning has now blossomed into a strong cultural foundation that draws from individual and collective memories. In fact, it has flourished so much that young children now host an annual school cultural ceremony based on the potlatch system. The Qqs Projects Society organizes a yearly Koeye Camp, an innovative Heiltsuk youth science and cultural camp program that takes place every summer in the Koeye River Valley. Through the camp, children learn, practise, and refine their Heiltsuk ways, including traditional dancing, singing, feasting, food harvesting, medicine making, and basket weaving. They participate at potlatches by performing the "Children's Play Dance Series." Because of a resurgence in Heiltsuk art and sewing regalia, most Heiltsuk now own a dance blanket and apron and have been bestowed with an ancestral name – practices that had earlier been banned during an era when Heiltsuk names were forcibly replaced by Christian ones derived from the Bible. Shared inheritance and cultural history that had been denied for so long were being rewoven through the strands of recollection. Through this, my Heiltsuk child started to grow again.

Local control of local education was assumed in 1976. As a founding member of the community school board, I was excited and filled with hope at the possibilities of Heiltsuk content enhancing Western curriculums, enabling our students to learn from the best of both worlds. For the first time in one hundred years, our young people didn't have to leave home to attend high school. We expected community members to flock to the school to support and celebrate the learning of their children. We were mystified when that didn't happen. Over time, we learned that many of our people had had negative educational experiences. Most adults had attended residential schools and lived with toxic memories that resulted in fear and mistrust of educational systems. To gain trust and support, we created family-focused events. Forty-two years later, the community school concept is a reality, and there are more than twenty-five certified Heiltsuk-born teachers who work here and elsewhere: a living example of how negative memories can be healed through patience, encouragement, and inclusion.

For many years, I've had a recurring nightmare of being lost. In this troubling dream, I try to find my way home but am unsuccessful. I find myself in alien settings, where I fear for my life, but I'm determined to go home. Last month, however, I had a dream in which I reached closure. In the dream, I walked down a long road that would lead me home. There were deep and dangerous potholes, a treacherously cracked pavement, and seemingly insurmountable barriers over which I had to climb to continue on my journey. At some point, I came upon an old man slumped in a wheelchair and clothed in rags. The wheelchair sat at a sharp angle, and the old man was at risk of tumbling off the road. I stopped to help and was shocked to discover that the man was my father. He, too, was trying to find his way home after escaping from a residential school. I pushed his wheelchair, and we went home. My mother was so happy to see us that she invited relatives to feast with us. In a short time, my father's age was reversed, and he was transformed into a younger, more vital, and strong man. Our relatives embraced us and celebrated our return home.

I woke from the dream with happy and sad tears streaming down my face, happy because we had found our way back home, sad because I now understood my father's experiences. He had been taken away to attend residential school at the age of seven and returned when he was sixteen years old. He passed away four years ago, and during our life together we had difficulty expressing our feelings for each other. He was hard-working, disciplined, responsible, and caring. He was the son of a hereditary chief who had provided for and protected his family. As our cultural ways grew, I encouraged him to take on his rightful role as a *Hemas* (hereditary chief). At first, he was reluctant, but I promised to help him based on what I had learned about our potlatch system. In the end, he did host a potlatch and ascended to his rightful place as a Hemas. He connected with his extended family from another village, and they came to participate in and support him in his cultural celebration. They remembered his paternal lineage and, through this, helped him to heal from that long-standing disconnection. This was such good medicine for his Heiltsuk soul.

As I continued my journey to my Heiltsuk self, I was blessed with many mentors. One of them was an old man who telephoned me often to share stories and educate me in Heiltsuk knowledge and teachings. To paraphrase, he always began our lessons by saying, "Put on your Heiltsuk mind and leave your white mind alone for a while, so I can teach you something about us." Or he would begin by asking me to look at my skin and tell him what colour it was on that day. He was a fluent Haíłzaqv speaker and had limited command of the English language. When we spoke, he would remind me to speak slower – not because he was stupid, but because he had to convert into his Haíłzaqv mind what I had said in English, to think about it in the Haíłzaqv context, change it back to English, and then respond. When I first heard him say this, I was completely taken aback because of the depth of teaching that he imparted. In essence, he taught me that the two worlds of which I was a part were very different and that both had value. He reminded me that I was born into our Heiltsuk world, had been removed from it, and then became confused and lost. His candor helped me to realize how much my mind had been acculturated and how mindful I must be to decolonize my indoctrinated Western thinking and reality. His Heiltsuk education helped me find my way home.

The ancestral name that I inherited from my maternal grandmother is Hilistis. It's a name that comes from an old village where an animal race circumnavigating the world took place. Whichever animal returned home first would signify the reigning animal clan of that village. In the story, the raven is first to return. Loosely translated, *Hilistis* means "starting out on a journey and staying on course until it is completed in full circle by returning home." What an appropriate name, given my quest to relocate my Heiltsuk self and find my journey home. The gift of memory has been instrumental in my Heiltsuk cultural reconnaissance. At last, the twelve-year-old Heiltsuk girl in me has become an old woman who, in turn, has become a knowledge holder and teacher. It is now her turn to share with others who are trying to find their way back to their cultural selves.

# ECOLOGICAL AMNESIA

*Wade Davis*

SOME YEARS AGO, I visited two places that in a different, more sensitive world would have surely been enshrined as memorials to the victims of the ecological catastrophes that occurred there. The first was the site of the last great nesting flock of passenger pigeons, a small stretch of woodland on the banks of the Green River near Bowling Green, Ohio. This story of extinction is well known. Yet until I stood in that cold, dark forest, I had never sensed the full weight, scale, and violence of the disaster.

At one time, passenger pigeons accounted for 40 percent of the entire bird population of North America. In 1870, when their numbers were already greatly diminished, a single column 1 mile wide and 320 miles long, containing an estimated 2 billion birds, passed over Cincinnati on the Ohio River. In 1813, as James Audubon travelled in a wagon from his home on the Ohio River to Louisville, some sixty miles away, a stream of passenger pigeons filled the sky, and the "light of the noonday sun was obscured as by an eclipse." He reached Louisville at sunset, and the birds continued to come. He estimated that the flock contained over 1 billion birds, and it was but one of several columns of pigeons that blackened the sky that day. Audubon visited roosting and nesting sites and found trees two feet in diameter broken off at the ground by the weight of birds. He saw dung so deep on the forest floor that he mistook it for snow. He compared the noise of the birds taking flight to that of a gale, the sound of their landing to thunder.

It is difficult now to imagine the ravages that would destroy this creature within fifty years. Throughout the nineteenth century, pigeon meat was the mainstay of the American diet, and merchants in eastern cities sold as many as eighteen thousand birds a day. Pigeon hunting was a full-time occupation for thousands of people. A typical shooting club would go through fifty thousand birds in a weekend competition. By 1896, there were only a quarter million birds left. In April of that year, the birds came together for one last nesting flock in the forest outside of Bowling Green. Telegraph wires hummed with the news, and the hunters converged. In a final orgy of slaughter, over two hundred thousand pigeons were killed, forty thousand were mutilated, and one hundred thousand chicks were destroyed. A mere five thousand birds survived. On March 24, 1900, the last passenger pigeon in the wild was shot by a young boy. On September 1, 1914, as the Battle of the Marne consumed the flower of European youth, the world's last passenger pigeon died in captivity.

When I left the scene of this final slaughter, I travelled west to Sioux City, Iowa. There, I was fortunate to visit a remnant patch of tall-grass prairie, a 180-acre preserve that represents one of the largest remaining vestiges of an ecosystem that once carpeted large swaths of North America. As I walked through that tired field, my thoughts drifted from the plants to the horizon. I tried to imagine buffalo moving through the grass, the physics of waves as millions of animals crossed the prairie.

As late as 1871, buffalo outnumbered people in North America. In that year, one could stand on a bluff in the Dakotas and see nothing but buffalo in every direction for thirty miles. Herds were so large that it took days for them to pass a single point. Wyatt Earp described one herd of a million animals stretching across a grazing area the size of Rhode Island. Within nine years of that sighting, buffalo had vanished from the plains.

The destruction of the buffalo resulted from a campaign of biological terrorism unparalleled in the history of the Americas. The policy of the federal government of the United States was explicit: exterminate the buffalo and destroy the commissary of the great cultures of the plains. Over 100 million bison were slaughtered. A decade after Native resistance collapsed, the general who orchestrated the campaign advised Congress to mint a commemorative medal with a dead buffalo on one side, a dead "Indian" on the other.

As I thought of this history, standing in that tall-grass prairie, what disturbed me most was the ease with which we have removed ourselves from this ecological tragedy. Today, the good and decent people of Iowa live contentedly in a landscape of cornfields claustrophobic in its monotony. The era of the tall-grass prairie, like the time of the buffalo, is as distant from their lives as the fall of Rome or the siege of Troy. Yet the terrible events unfolded but a century ago, well within the lifetime of their grandparents. This capacity to forget, this fluidity of memory, has dire implications in a world dense with people, all desperate to satisfy their immediate material needs. Confronted by the consequences of our actions, there is always the path of forgetfulness.

Humans, of course, have long impacted their environments. Pre-Columbian peoples deforested much of Andean Peru long before the rise of the Inca. The severely eroded and barren hills of the Loess Plateau were once a flat and fertile plain covered with forests and rich grasslands, the cradle of ancient Chinese civilization. Romans and Greeks over many centuries destroyed the rich forests of Lebanon and virtually all timbered lands surrounding the Mediterranean. The extent of deforestation caused by successive Mayan civilizations in the lowlands of the Petén is only beginning to be fully appreciated. Successive generations of Polynesians exhausted the resources of Rapanui, or Easter Island, literally eating themselves out of house and home.

The story of Easter Island has become an ecological fable because it speaks directly to the fate of the world today. Yet we remain haunted by a capacity to forget that lingers like a vestigial and necrotic appendage on the body of humanity. Perhaps ecological amnesia served our needs in the past, as we gradually degraded the natural world over generations. But today the time frame has contracted dramatically, even as our capacity to destroy the environment has expanded to an industrial scale, with no place on the planet beyond reach. If the Mediterranean forests fell to the Roman axe over centuries, the landscape of Sarawak, homeland of scores of Indigenous cultures dependent on the forest for their survival, was converted by chainsaw and bulldozer to wasteland in a mere generation. And yet we continue to forget.

How many of us remember, for example, that as recently as the 1920s the Colorado River delta was lush and fertile, a "milk and honey wilderness," in the words of Aldo Leopold. Today it is a wasteland of barren mudflats, with the river but a toxic trickle in the sand. The Gulf of Alaska once turned a golden hue with the sheer numbers of returning salmon, a sight unlikely to be seen again. Off the shores of Newfoundland, cod were so abundant that ships with wind in their sails made little progress, blocked by the density of fish in the water. Europe and much of the New World lived on the catch for three hundred years. Then, in the years of my youth, factory ships industrialized the fishery and in a single generation reduced the species to a shadow in the sea.

As recently as the 1920s, Haiti was 80 percent forested. Today, less than 2 percent of its forest cover remains. I recall one day walking along a barren ridge with an old man who waxed eloquent as if words alone might squeeze beauty from the desolate valley of scrub and half-hearted trees that reached before us to the horizon. Though witness to an ecological holocaust that had devastated his entire country within a century, he had managed to adorn his life with his imagination. This capacity was inspiring but also terrifying. People appear to be able to tolerate and adapt to almost any degree of environmental degradation. As the farmers of Iowa today live without wild things, Newfoundlanders survive without cod, and the people of Haiti scratch a living from soil that may never again know the comfort of shade.

From a distance, both in time and in space, we can perceive these terrible and poignant events for what they were, unmitigated ecological disasters that robbed us and the future of something unimaginably precious in order to satisfy the immediate mundane needs of the present. The luxury of hindsight, however, does little to cure the blindness with which we today overlook deeds of equal magnitude and folly. In a manner that will be difficult for our descendants to comprehend, we drift towards a world in which people take for granted an impoverished environment, transformed by foolish negligence and reduced by expediency to a shadow of the glory that once was. In three generations, a mere moment in the history of our species, we have throughout the world contaminated the water, air, and soil, driven countless species to extinction, dammed the rivers, poisoned the rain, and torn down the ancient forests. As Harvard biologist E.O. Wilson reminds us, this era will not be remembered for its wars or technological advances but as the time when men and women stood by and either passively endorsed or actively supported the massive destruction of biological diversity on the planet.

Given the dire consequences, how might we explain this peculiar and ultimately self-destructive capacity to shed memory and shift our expectations as we adapt to an increasingly impoverished world? Were this to be a fundamental adaptive trait of our species, we would surely find evidence scattered throughout the ethnographic record. But most assuredly we do not. Indeed, to the contrary, most Indigenous peoples cultivate fidelity to the deepest of memories, myths that both link the living to the ancestral past and illuminate the way to the future. Take, for example, the Indigenous peoples of Australia, who thrived as guardians of their world for over fifty-five thousand years. In all that time, the desire to improve upon the natural world, to tame the rhythm of the wild, never touched them. Indigenous people accepted life as it was, a cosmological whole, the unchanging creation of the first dawn, when the primordial ancestors sang the world into existence.

The Europeans who washed ashore on the beaches of Australia in the last years of the eighteenth century lacked the language or imagination even to begin to understand the profound intellectual and spiritual achievements of Indigenous Australians. What they saw was a people who lived simply, people with modest technological achievements, strange faces, and incomprehensible habits. To European eyes, the Indigenous people were the embodiment of savagery. By the early years of the twentieth century, a combination of disease, exploitation, and murder had reduced the Indigenous population from well over a million at the time of European contact to a mere thirty thousand. In little over a century, a land bound together by songlines – through which the people moved effortlessly from one dimension to the next, from the future to the past and from the past to the present – was transformed from Eden to Armageddon. The manner by which the Indigenous peoples of Australia imbued the natural world with a sense of the sacred is not contrary to science but rather an acknowledgment of the complexity and wonder of ecological and biological systems that science illuminates. It suggests that our capacity to forget and adapt to successive degrees of environmental degradation is less a human trait than a consequence of culture.

As a young man, I was raised on the coast of British Columbia to believe that the rainforests existed to be cut. This was the ideology of scientific forestry that I studied in school and practised in the woods as a logger. This cultural perspective was profoundly different from that of the local First Nations. Whereas I was sent into the forest to cut it down, a Kwakwa̱ka̱’wakw youth of similar age was traditionally dispatched during his *hamatsa* initiation into those same forests to confront Huxwhukw and the Crooked Beak of Heaven, cannibal spirits living at the north end of the world. Is the forest mere cellulose and board feet? Or is the forest the domain of the spirits?

Herein, perhaps, lies the essence of the relationship between many Indigenous peoples and the natural world. The malarial swamps of New Guinea, the chill winds of Tibet, and the white heat of the Sahara leave little room for sentiment. Nostalgia is not a trait commonly associated with the Inuit. Nomadic hunters and gatherers in Borneo have no conscious sense of stewardship for mountain forests that they lack the technical capacity to destroy. What these cultures have done, however, is to forge through time and ritual a traditional mystique of the earth that is based not only on deep attachment to the land but also on far more subtle intuition – the idea that the land itself is breathed into being by human consciousness. They do not perceive mountains, rivers, and forests as being inanimate, as mere props on a stage upon which the human drama unfolds. For these societies, the land is alive, a dynamic force to be embraced and transformed by the human imagination, sustained by memory.

Perhaps this explains why it is impossible for the Haida to forget the colour of the sea in the fall, and why the Lakota still hear the thunder of bison crossing the prairies, and why, in the wasted homeland of the Penan in Borneo, shadows still mark the ground where trees once stood in the forest. Just as eighteenth-century slavers concocted racial fantasies to mask the evil of their trade, perhaps we have learned to shed memory to avoid confronting the actual consequences of our egregious violations of the natural world. Our shifting expectations and dimming memory are less an adaptive trait than a reflexive impulse. If we are responsible for the numbing of our own senses, we can surely awaken to new possibilities as stewards of life, inspired by Indigenous peoples who have walked this path before us, guided by a conscience informed by memory.

# CLIMATE TALES

*Edouard Bard*

CURRENT AND FUTURE climate change necessarily preoccupies much of our attention. But human perception of climatic variation is imperfect and unreliable. Extreme events such as summer heat waves, abnormally cold winters, and major storms loom large in our memories. Our perception and recall of slow environmental change are less precise. As individuals, we have a limited capacity to perceive thc one degree Celsius of global warming that has taken place over the last century, nor can we appreciate the significance of this change for global climate and ecosystems. Limitations in our perspective help explain our inertia when it comes to mitigating and adapting to a shifting climate. To understand the potential impacts of ongoing environmental change, and to imagine our potential future, we must turn to the physical, nonhuman memories of past climates. Written in the oceans, land, and ice sheets of our planet, these archives bear silent witness to the massive environmental changes that have occurred on Earth over geological time.

Earth's climate is shaped by complex interactions among the atmosphere, oceans, land, and ice, which are governed by the fundamental laws of physics and chemistry. These components of the climate system operate on vastly different time scales. For instance, gases, liquids, and solids have different viscosities, meaning that air, water masses, and glaciers move at different rates. Molecules in the atmosphere can be transported over thousands of kilometres in the span of days, whereas an ice sheet will creep forward only a few metres per year. Chemical processes likewise occur at different rates. The reaction time of aerosol formation can be less than seconds, whereas the chemical weathering of silicate rocks and the dissolution of deep-sea minerals, both critical to the uptake of atmospheric carbon dioxide, stretch out over millennia. These different physical and chemical processes occur simultaneously, leading to a complex system in which long-term and short-term events can be recorded at the same time. We can look to the geological past to see evidence of massive glaciations that occurred over tens of thousands of years. We can also see evidence of shorter-term climatic events associated with the eruption of a volcano and the release of ash plumes into the stratosphere. Much as in our own human memory, these long- and short-term memory traces can interact with one another in complex and sometimes surprising ways. Consider, for example, short-term variations in weather. A specific weather event at a particular location will not be "remembered" explicitly by the climate system for more than a few weeks. Owing to atmospheric turbulence and mixing, air masses quickly lose the characteristic signature of previous physical states, leaving no detectable trace of past weather conditions.

Devastating winds associated with a hurricane, fuelled by an oceanic heat source, dissipate quickly as the air mass moves over land. Within several weeks, we can no longer accurately trace the storm's trajectory or reconstruct its prior temperature, wind speed, or other properties. The destruction brought on by a storm will thus last far longer than that storm's physical presence within the climate system. We can understand longer-term weather patterns from the statistical average of many individual events, but predicating a single weather perturbation eludes us. For this reason, we can make projections about the average frequency and intensity of hurricanes on a warming planet over years and decades, but we can't accurately forecast the trajectory and characteristics of a storm more than a few weeks into the future.

Earth's climate system may have a fickle short-term memory, but its long-term climate history has been recorded and stored in multiple ways. During the last glacial period, about twenty thousand years ago, Earth's climate bore little resemblance to what we experience today. Average global temperatures were five degrees Celsius colder, and enormous ice sheets covered Canada and Scandinavia. These large ice masses, up to four kilometres thick in some places, modified the shape of the planet as they pressed down on Earth's surface with their enormous weight. After the glaciers retreated, about ten thousand years ago, Earth's surface began to rebound back towards its original shape, much like a foam mattress after the weight of a body has been removed from it. This isostatic rebound is ongoing today and can be measured accurately, with rates exceeding one centimetre per year in extreme cases. This process helps explain the complex patterns observed in sea level changes across the planet. There is no doubt that global average sea level is now rising approximately three millimetres per year because of global warming, which causes the thermal expansion of sea water and the melting of mountain glaciers and ice sheets in Greenland and Antarctica. But we can't understand the actual patterns of sea level change in any particular location without taking into account traces of Earth's history as a glaciated planet. For example, the sea level along the US East Coast has risen significantly faster than the global average, while other regions in the tropical Indian and Pacific Oceans exhibit lower than average rates of sea level rise. In areas close to former ice sheets, including Hudson Bay and the Baltic Sea, sea levels are actually decreasing because of the isostatic rebound of the land surface, providing a detectable physical memory of past glaciation.

Earth retains memories of its distant past in more direct ways. Atmospheric temperature variations affect surface ocean temperatures, which are then transferred into the deep sea by the vertical mixing of water masses and large-scale ocean currents. It can take up to two millennia for surface water to reach the bottom of the north Pacific, and over this long period the temperature of a water mass will remain largely unchanged. This means that the present-day temperature of the deep Pacific Ocean reflects atmospheric conditions during the early days of the Roman Empire. Ancient ice sheets likewise hold memories of past atmospheric temperatures. Precise measurements in the Greenland ice sheet have shown that evidence of the cold atmospheric temperatures of the last glacial period still resides about two kilometres beneath the surface. This temperature memory is slowly diffusing away but, assuming that climate stays stable, it will take tens of millennia for the ice at these depths to reach thermal equilibrium with the overlying atmosphere.

The history of Earth's climate is thus written all around us, in deep ocean waters, in sediments, and in glaciers. The challenge is learning how to read this memory, to understand its implications for the present and to preserve the record for the future. Unfortunately, we haven't been collecting observations for long. Atmospheric temperature measurements go back only to the eighteenth century. Data for oceanic temperatures and sea level reach back no farther than the nineteenth century. And these data are strongly biased towards Europe and North America. In some cases, scientific measurements can be supplemented by Indigenous oral histories, but they too are geographically limited and only go back about ten thousand years, at best. We need a longer history.

To go back farther in time, and to improve the geographic range of observations, climate scientists have developed paleoclimatic indicators – so-called climate proxies – that can be measured in a wide variety of natural archives, including glaciers and ice sheets, groundwater, sediments, stalagmites, fossil corals, and ancient trees. To establish a link between a proxy measurement and a climate fact, scientists apply what they know about chemical thermodynamics or the biological adaptations of various organisms, whose remains may be preserved in the geological archive. The science of geochemistry and biology can help us interpret the ancient signatures we find in various Earth system archives. Admittedly less precise than instrumental data, climate proxies can nevertheless generate valuable information about the past when applied with care. They are the only means at our disposal to recover historical climate features and examine the Earth system's deep memory.

Like other forms of memory, the information held in climate proxies can disappear or be distorted over geologic time. Paleoclimate studies must therefore be highly redundant, and scientists typically measure several proxies in the same archive and use different archives to reconstruct a local climate record. Like a jury in a criminal trial, we must arrive at our best guess of the truth by reconstructing multiple lines of evidence from different sources and sites. As an example, consider the use of isotopic thermometers in polar ice caps, which are based on measuring small changes in the relative abundance of different isotopic forms of oxygen and hydrogen in the water molecules of different ice layers. These signatures reflect atmospheric temperatures around the time when snow accumulated on the ice cap, building up each successive layer of ice laid down in a chronological sequence.

In the early 1990s, Europeans and Americans raced to drill adjacent sites on the summit of the Greenland ice sheet. Within the top three kilometres of the ice core, the two isotopic temperature records were nearly identical, showing abrupt and repeated climate shifts over the past hundred thousand years of Earth history. Below this depth, however, the two climate records, each from a slightly different site, diverged significantly, particularly in the oldest sections of ice. In hindsight, we know that these disparities reflected local melting of the deep ice, influenced by geothermal heating, that led to complex folding and mixing within the ice sheet. This physical distortion of the ice core record significantly complicates how we interpret the various proxies and must be taken into account if we are to arrive at a correct understanding of past climate history. And other examples of memory distortion reside in various climate archives. Sediments can be mixed, homogenized, and transported, disturbing the natural layering that we use to infer a chronology; minerals can be dissolved and then reformed, altering the chemical composition from which we derive the climate record.

An extreme example of geological memory loss and distortion has unfolded over the past few decades. The ice of low and mid-latitude mountain glaciers, such as those on Mount Kilimanjaro in Tanzania and in the South American Andes, houses a significant repository of climate memory. In the last half of the twentieth century, these glaciers began to rapidly disappear, and it's highly probable that most will be gone before the end of this century, taking with them the climate history encapsulated in their frozen archives. The international Ice Memory project, launched by French and Italian scientists in March 2017, aims to create the world's first library of archived glacier ice as a "scientific heritage for generations to come." The project has already collected ice cores on Mont Blanc in the French Alps and on Mount Illimani in the Bolivian Andes. These cores will be shipped to central Antarctica, where they can be maintained indefinitely, frozen without electricity or other significant infrastructure, thereby creating an archive of Earth's climate history.

Science can help us supplement, store, and restore our failing memories about Earth's climate. It allows us to identify the climatic causes of events that affect human societies such as food shortages, large-scale epidemics, or even wars. We need to document and understand past environmental changes as they are remembered in the materials of the Earth, working against our human tendency to either not notice or simply forget.

# MAKING RUINS

*Sabine Bitter and Helmut Weber*

ARCHITECTURE CAN CARRY MEANING, hold memories, and make history. This capacity can be expressed on a small scale, such as in the representation of buildings on a coin or a banknote. It can also occur on grander scales that influence the construction and preservation of particular urban histories. Such histories live most tangibly at the scale of the city, but they are also expressed at a national level, where architecture can be deployed to reinforce or to challenge collective social memory. While architecture holds intrinsic meanings and memory, it can also be reworked to project a vastly different historical narrative than the one it originally represented. What happens, then, to architecture's capacity to constitute a collective memory when layers of a newly constructed Grand History enclose and cover up the original material body of a built structure? Once obscured, how are we to reactivate architecture's capacity to hold lived history?

These questions were sparked by the unusual urban situation of Skopje, the capital city of the Republic of Macedonia. Over the last

The first five photographs featured in this essay were taken by Sabine Bitter during her research in Skopje in 2016.

few years, the federal government has forcefully reshaped this city, through the "Skopje 2014" program, into a historical city with ideals rooted in antiquity. This refabrication has been embodied through the construction of monuments and zombie-like buildings aimed at erasing all traces of the city's socialist history. A sketch of the last five decades of Skopje's history reveals what is at stake for the city and why questions about the instrumental use of architecture are intrinsically tied to questions of social and collective memory.

To understand the current context, one must understand the modern history of Skopje. On July 26, 1963, a massive earthquake hit the city, resulting in the loss of close to one thousand lives and 120,000 homes. About two-thirds of the city's built space collapsed in ruins. Despite ongoing geopolitical tensions at the height of the Cold War, the United Nations launched an unprecedented international act of solidarity to provide immediate help, shelter, and resources for the city, then part of the Socialist Federal Republic of Yugoslavia. At a time when all contact points between East and West Berlin were closed, Skopje was declared an open city. Yugoslavia's "president for life," Josip Broz Tito, welcomed the support of the United Nations and renamed Skopje the "city of solidarity," a symbol of international assistance that would build bridges across the borders of nation-states and the boundaries of ideological systems.

Within a year, these international efforts gave rise to a plan to construct a completely new city. In February 1965, the United Nations Special Fund and the Yugoslav government, in cooperation with the International Union of Architects, invited four Yugoslavian and four international firms to participate in the "Skopje City Center Master

Plan" competition. The winning prize for the "Open City of Skopje" was split sixty-forty between the team of Japanese architect Kenzō Tange and Zagreb-based firm Miščević and Wenzler. The Skopje Institute for Town Planning and Architecture subsequently joined this unique international collaboration, which was dominated by Tange's dynamic vision but included iconic buildings by the Macedonian architect Janko Konstantinov. At that point, Tange was internationally renowned for his Hiroshima Peace Center, built in 1955, and for his experimental Tokyo Bay plan, proposed in 1960, which incorporated ideas of organic biological growth and a belief in technological progress into architectural design. These ideas would eventually grow into the Metabolist Movement, within which Tange emerged as a key figure. He and his followers understood civic architecture as a flexible bridge between rapid technological development and the social needs of humanity. Although Tange's master plan for Skope was never realized in its entirety, some of the buildings such as his train station and housing blocks of the City Wall (1966, 1965–68) and Janko Konstantinov's Post and Telecommunications Building (1970–86) were completed, adding examples of international modernism and local brutalism to Skopje's post-earthquake Open City Center. Skopje, though incomplete in its modernist grand vision and in the master plan, was a test case of the internationalism of modernism, ranging across different regional contexts, local spatial knowledges, and lingering national discourses.

In the three decades since elements of Tange's plan have been implemented, geopolitical shifts, the emergence of the European Union, and localized nationalistic struggles have put pressure on Macedonia and Skopje. Since its declaration of independence from

the former Yugoslavia in 1991, Macedonia has had to come to terms with its new identity as a democratic state. The country was admitted to the United Nations in 1993 under the provisional description of the Former Yugoslav Republic of Macedonia. Greece has a province named Makedonia within its own territory, and the conflict over the use of the name *Macedonia* continues to this day. This struggle for naming is symptomatic of the recent wave of emerging nationalism within the European Union. Similar to other states that have recently emerged from the former Yugoslavia and Soviet Union, Macedonia faces a surge in ethnically based nationalisms and religious identities. While Skopje was never subjected to a civil war on its territory (unlike cities such as Sarajevo), the force of emergent nationalisms has nevertheless shaped its city spaces.

In an unexpected turn towards Western neoliberal democracy, Macedonia elected a conservative nationalist government in 2006. This political shift was underpinned by aspirations to join the European Union and create a post-Yugoslavian national identity. It was in this context that the "Skopje 2014" program was launched in 2010. Seen in this light, "Skopje 2014" can be understood as a constructed portrayal of a false Macedonian history, pointedly and intentionally obfuscating traces of the country's collective socialist past in favour of a hero narrative rooted in the era of Alexander the Great, who ruled the ancient Greek kingdom of Macedon in the third century BC. The central city square is dominated by recently created monuments to Alexander, scaled to overshadow those structures that are reminders of Macedonia's recent socialist past and Yugoslavia's historical position in the Movement of Non-Aligned Countries – a movement that stressed independence and decolonization. Importantly, also lost in the construction of a new national and urban narrative is the unique moment of local modernity and international solidarity represented by Tange's post-earthquake city design and buildings.

Tange's modernist architecture, and the concrete brutalist buildings of his Yugoslav contemporary Janko Konstantinov, haven't simply been destroyed, erased, or replaced. In a strange program of reverse facadism – a practice that preserves only the facade of a historical building – the exteriors of these historically significant modernist and brutalist masterpieces have been disguised by a thick layer of baroque architectural elements and columns manufactured

from polystyrene foam and cheap plastic. Although these architectural stage props are meant to represent antiquity, they are fabricated with contemporary materials that ironically deprive the architecture of its capacity to be a carrier of authentic memory, even as that architecture proclaims historical remembrance. This destructive make-over process is a form of imposed cultural amnesia, leaving no trace or material rubble to be commemorated.

How are we to understand the conflict and contradictions produced when the existing cultural memory of a city is covered over by a false narrative constructed from an imagined antiquity? Indeed, the "Skopje 2014" program has been met with significant controversy. Citizens, architects, artists, activists, and intellectuals have criticized the program in local and international media and have initiated a "colorful revolution" in which activists shoot paint balls or throw plastic bags filled with brightly coloured paint at the monuments and buildings central to the new constructed narrative of Macedonian history.

Can art enable citizens to remember another past, to recall the body of architecture evaporated or absorbed into those freshly fabricated "historical" buildings? The photograph featured opposite is part of an artistic work we have created called *Making Ruins*. This work joins other forms of protest in questioning this thick layer of fictitious history. The photographs of the last remains of Tange's and Konstantinov's buildings point to the corpses of architecture hidden away under foamy camouflage. The artistic intervention is meant to rework the still visible architectural remains into ruins, acting to interrupt the linearity of false historization, preserving memories of a particular moment of past international solidarity and claiming an alternative future for Skopje.

Installation view of *Making Ruins*, Sabine Bitter/Helmut Weber, MuseumsQuartier Vienna, 2016. Photo by Suchart Wannaset.

ACKNOWLEDGEMENTS

We would like to thank Suzana Milevska, Sašo Stanojkoviḱ, and Jeff Derksen for generously sharing their knowledge.

# TIMOTHY FINDLEY'S *THE WARS*

*Sherrill Grace*

DURING THE LATTER HALF of the twentieth century and in response to critical reflection on the two world wars, cultural studies experienced a so-called memory boom that prompted scholars to analyze memory as a process of forgetting and recovering memories.[1] For Canadian novelist Timothy Findley (1930–2002), memory was a form of narrative constructed by public institutions and official history with a capital H, by artists in their creative works, and by the stories we tell about who we are. For Findley, the trauma of the First World War persisted in his family, and he recalled their war stories fifty years later under the shadows of the Second World War and the war in Vietnam.

His novel *The Wars*, published in 1977, is one of the earliest and most influential of Canadian narratives about memory and war. It prepared the way for increased literary attention to the First World War and for new ways of conceptualizing memory. In it, Findley situates one young officer's life within the shattered landscapes of the Western Front and the ruptured lives of his family at home. Second Lieutenant

Robert Ross may be the hero (or antihero) of the story, but the key character is the biographer who conducts research to reconstruct Robert's story. To read *The Wars* is to participate in the complex, never complete process of rediscovering a past that is unknown to us but is nonetheless ours. As we read, we become haunted by the war, much as the biographer is. "You begin at the archives," Findley writes, before warning, "As the past moves under your fingertips, part of it crumbles. Other parts, you know you'll never find."

*The Wars* is a Canadian classic. It won a Governor General's Award in 1977, has never been out of print, and has been taught for decades in Canadian high schools and universities. Widely translated, adapted for film and stage, it represents Canadian cultural memory of the war abroad. The power of the story stems from the quality of Findley's writing and from its sympathetic subject. Robert Ross is a gentle, sensitive young man, hardly the type to choose a military career, but social and family pressure leads him to volunteer for Canada's citizen army, and he leaves for the front at the age of nineteen. Robert learns to kill because he must do his duty and protect the soldiers under him; however, his experiences drive him to rebel against his senior officers, to shoot a captain whose orders he deems insane, and to escape the scene of carnage and fire at Bailleul in June 1916 in an attempt to save a group of horses from the German barrage. When he is found, resting in a barn with the horses, the military police set fire to the barn to force him out. He is captured but is too badly injured to appear before the military tribunal. He is found guilty *in absentia*. Left in the convalescent home on the D'Orsay family's English estate, he dies in 1922, shortly before his twenty-sixth birthday.

The story of Robert's life is told through three categories of memory and with three types of narrator who remember him. The German scholars Astrid Erll and Ansgar Nünning provide useful terms for identifying these categories of memory.[2] Their first category, the memory of literature, pervades the novel through references to other texts, such as British war poetry and Carl von Clausewitz's military treatise, *On War* (*Vom Kriege*). This form of memory is also used to evoke material objects that exist outside the text, such as war art. The second, what they call literature as a medium of collective memory, serves to contextualize Robert's story within Canadians' collective memory of the war as produced by historical facts and archival data. The third, memory *in* literature, operates within the novel as personal, private acts of remembering and forgetting, which often conflict with collective memory and with official history, which claims Canadian national identity was forged in the First World War.

The unnamed biographer, who is searching for the truth about Robert's life, faces many obstacles in his research: the archives are incomplete; Robert's letters home leave much unsaid; court documents are available but report only the military view of Robert's actions; eyewitness testimony is full of contradictions; many old veterans who knew Robert can no longer recall events; and others refuse to be interviewed. There are, however, two elderly women who knew Robert and share their memories of him with the biographer, who includes transcripts of their interviews in the story. These women provide personal memories, which often contradict the official public story. Outside the biographer's narrative is a traditional third-person narrator, who provides information not accessible to the biographer.

Allusions to and quotations from British war poets such as Siegfried Sassoon and Wilfred Owen are fairly common in novels by writers of later generations who must re-create memories of the First World War. In *The Wars*, allusions to G.A. Henty's *With Wolfe in Canada* and to Clausewitz's *On War* highlight Findley's ironic remembering of the war. Even if we have not read Clausewitz, his name alone has entered the lexicon of cultural memory, and it emphasizes how misguided his military strategies were for a twentieth-century war. One especially effective trigger for Canadian cultural memory of war is the biographer's description of Benjamin West's painting *The Death of General Wolfe* (1770). The biographer associates this painting with a photograph of Robert in uniform that occupies a place of honour in his parents' living room. This photograph represents the seriousness of death, but it also conveys an ironic image of death as a glorified, sanitized, Christ-like sacrifice.

Such familiar rhetoric and images deceive us about the reality of war and death, but they are deeply embedded in our official narrative that tells us Canada came of age in this war. To further prompt our collective memory of the war, Findley draws extensively on military facts by dating sections of the story as if the fictional biographer were writing a diary, which we read. This helps us locate Robert in key battles at actual sites in Flanders. For example, after a heavy barrage at St. Eloi in February 1916, Robert is trapped with his men in a crater during a gas attack. The biographer counts off the hours for the soldiers, who, without functioning gas masks, must urinate on their shirttails, which they hold over their faces, and burrow into the mud until the gas dissipates. This kind of specific information authenticates Robert's story by securing the fiction to the facts of history and geography. Insofar as such facts are common knowledge – recorded in history books, films, published memoirs, and war art exhibitions or familiar from battle site tours and war cemeteries – they foster a remembered narrative of the war that accommodates Robert's story and blurs the distinction between fact and fiction.

What readers of *The Wars* won't know is that Findley's inspiration for his novel came, in part, from his uncle's First World War journal. Timothy Findley inherited this journal, which contains a fading photograph of Thomas Irving Findley (1895–1933) in uniform and transcribed letters with observations that his uncle sent home during the war. These personal stories inspired Findley to use details of people and scenes described in the journal (itself a record of events remembered), and he adapted its sequencing of dates and place names for his fictional narrative, thereby transforming personal memory into public cultural memory for the novel. By underpinning the main narrative with facts and references external to the novel (for example, newspaper reports, court documents, paintings, and others' writing), this personal connection with the past helped Findley bring the war and its fictional characters alive in a reader's present.

The third category of memory in *The Wars* is the private act of remembering and forgetting that supplements and, at times, resists official collective memory. The transcripts of Marian Turner, the Canadian nurse who "received" Robert's badly burned body, and of Lady Juliet d'Orsay, who cared for him until he died, provide personal accounts charged with painful memories and considerable anger at the war in general and at Robert's treatment by the military. Turner recalls him as "an homme unique," characterized by integrity and bravery; because she cannot forget, she confesses to the biographer that she had offered to spare Robert his agony with an overdose of morphine, which he refused. Juliet d'Orsay reads sections about Robert from her 1914–16 diaries into the biographer's tape recorder, and she warns the biographer and the reader not to judge a man like Robert who was caught in appalling circumstances that we have not experienced. Memory enables both women to construct an alternative narrative to the official one of noble sacrifice and patriotism or, in Robert's case, of insubordination and murder. The war was, in their view, grotesque and insane, while men like Robert were sane and heroic.

An additional component of memory in *The Wars*, not discussed by Erll and Nünning, merges public, collective memory with private, individual memory. Findley invokes archives as repositories of memory at several points in the novel: we "begin at the archives" with many boxes of material, from maps and newspaper clippings to personal letters and snapshots, and return to this material at various points, most significantly at the end. As the library closes, the biographer arranges his research "in bundles–letters–photos–telegrams" but pauses to study a photograph of Robert with his sister taken before the war, when he was happy and alive. He turns the picture over and reads what is written on the back: "Look! You can see our breath." These words are life-affirming and are shared with us. Having read the novel, we can see this young man in our imaginations because he has been brought back to life through the operation of memory. We understand what society lost with his pointless death, and despite the lapse of time, we recognize in this fictional character our shared humanity. The institutional archive has yielded up a profoundly personal spur to memory and imagination.

*The Wars* is an exemplary work for exploring Canadian cultural memory of the First World War, for challenging official war narratives, and for valuing personal memories of traumatic events. In these ways it was ahead of its time. Today, exhibitions at the Canadian War Museum and works such as R.H. Thomson's multimedia memorial project *The World Remembers* show how the First World War haunts us and enjoin us to remember its lessons. Finally, with the plural title of his novel, Findley reminds us of the phrase "ripe for the wars" and of the persistence of war in our lives. He warns us that by forgetting war, we perpetuate its violence and destruction.

NOTES

1 See Andreas Huyssens on the memory boom in *Twilight Memories: Marking Time in a Culture of Amnesia* (London: Routledge, 1995) and Sherrill Grace, *Landscapes of War and Memory: The Two World Wars in Canadian Literature and the Arts, 1977–2007* (Edmonton: University of Alberta Press, 2014).

2 Astrid Erll and Ansgar Nünning, "Concepts and Methods for the Study of Literature and/As Cultural Memory," in *Literature and Memory: Theoretical Paradigms, Genres, Functions*, ed. Ansgar Nünning, Marion Gymnich, and Roy Sommer (Tübingen: Francke Verlag, 2006), 11–25.

# ECHOES ACROSS GENERATIONS

*Patricia M. Schulte and Judith G. Hall*

TRAUMATIC EVENTS can lead to vivid and lasting memories that deeply affect the lives of individuals and the collective consciousness of a society. These deep-seated memories can be passed across generations to shape families and entire cultures. But where do these memories reside? Does memory live solely in our brains, formed within our conscious and unconscious minds? Or might memory be a broader phenomenon? Recent scientific discoveries suggest the presence of biological memory carried in the very cells of our body, which can profoundly affect our health, our personality, and even our success in life.

DNA carries the genetic blueprints to make all of the proteins in our bodies. This DNA has been passed down across many generations, shaped by evolutionary processes selecting genes that enhance an organism's ability to survive and reproduce. In this way, our DNA sequences act as a type of biological memory that records the past evolution of our species. New observations, including epidemiologic and biological studies, suggest there is another type of biological memory that acts through so-called epigenetic mechanisms (*epi* = on top of) that control cell gene expression in specific ways. The mechanisms of this newly recognized biological memory have not been entirely worked out, and it is likely that there are multiple processes at work. It is now becoming clear, however, that epigenetic control of gene expression, in addition to the DNA sequences themselves, acts as a form of biological memory.

The terrible famine known as the Dutch Hongerwinter (Hunger Winter) provides an example of this biological memory in action. During 1944–45, late in the Second World War, the Nazis blocked all shipments of food and fuel into the western parts of occupied Holland. Compounding the privations of German occupation, the winter that year was particularly cold and harsh, and it followed a poor crop harvest in the fall. The Germans confiscated much of what food was available, and the Dutch people were reduced to eating anything they could get, including their prized tulip bulbs. That winter, more than twenty thousand people starved to death. The famine was ended only by the liberation of Holland in 1945.

The Dutch Hongerwinter was a tragedy of enormous proportions, but it is also a scientifically important example of a rare phenomenon: a famine in a well-educated, highly literate, and developed country that maintains detailed health care records. The survivors experienced a single well-defined period of famine followed by access to plentiful food during the postwar years. This makes the Hongerwinter a kind of natural experiment that provides an opportunity to study the long-term effects of early trauma (both psychological and physiological) on health outcomes in humans throughout life.

One of the first observations from studies of the Hongerwinter was that babies born to parents who lived through the famine had lower birth weights than normal, but only if their mother had experienced hunger during the later parts of her pregnancy. Conversely, babies whose mothers were starved only at the beginning of their pregnancy had normal birth weights. As doctors followed the health outcome of these babies, they found that those who had been born small stayed small throughout life. These smaller babies also tended to have very low rates of adult obesity compared to the general population. In contrast, babies whose mothers had experienced famine only early in gestation had normal birth weights but significantly higher rates of adult obesity, cardiovascular disease, and diabetes. As older adults, these babies are now exhibiting higher rates of dementia at earlier ages. Taken together, these observations suggest that a few months of famine very early during development can have effects that persist throughout life.

We are now several generations beyond the end of the Second World War, and the women and men who experienced the Hongerwinter now have grandchildren who have never experienced famine. But these grandchildren also show different health outcomes than people whose grandparents did not experience the famine. These data suggest the existence of a long-lasting multigeneration biological memory that can profoundly affect our health.

Observational studies in humans are notoriously difficult to interpret because of the many confounding social and biological factors that can influence health outcomes. To help disentangle the potential effects of these factors, scientists have conducted controlled experiments in many types of animals (ranging from laboratory mice to fruit flies). For example, in stickleback fish, exposing either a male or female to simulated attacks by a predator (using clay models) causes changes in the behaviour of their offspring. The offspring of "predator exposed" fish tend to react more strongly to predator cues and tend to be less willing to venture out of hiding places than the offspring of fish that have not been exposed to predators. As a result, these timid fish tend to do better in environments where predation risk is high, whereas bolder fish may do better in habitats with few predators, as they are more willing to venture out to forage for food. These results suggest that the effects of biological memory may be context dependent. Across many species, similar multigenerational effects of early life experiences have been documented, clearly showing that biological memory is a real and heritable phenomenon that occurs among many (if not all) animals. This biological memory is a record of past experiences and, potentially, those of our parents and grandparents. It affects the physiology and growth of our cells, potentially pre-programing offspring to deal with the conditions experienced by their parents.

We are still in the early stages of understanding the molecular mechanisms of biological memory, but studies on this subject are accumulating rapidly. The more we learn about developmental biology and molecular genetics, the more we realize that there are multiple types of environmental information being passed along throughout an individuals' life, and even across generations without changing the actual gene sequences.

Some forms of biological memory may be carried by specific cells or cell types. During pregnancy, cells are exchanged between the mother and the developing fetus. The resulting mixture of maternal and fetal cells, a microchimerism, is aptly named after the Chimera, a creature from ancient Greek myth composed of parts of more than one animal. These “foreign” cells migrate through the body, making their way to many organs, where they play complex roles in health and disease. These cells can persist in both mother and child for many decades, and even cells from miscarried pregnancies can remain in a mother’s body throughout her life. For example, a study of women who died late in life of natural causes found that almost 50 percent had male DNA in their brains, presumably as a result of cells transferred from male offspring during pregnancy. There can even be transfer of cells from older siblings to the mother and then on to younger siblings, or from the grandmother to the mother and then to the mother’s children. Families are thus tied together by a complex web of shared cells across generations.

The presence of embryonic or fetal cells in a mother’s body may help her tissues with healing and immune responses. For example, when the heart is injured following a heart attack, the fetal cells may migrate to the site of injury and differentiate into heart cells to help repair the damage. On the other hand, there is also evidence to suggest that the presence of fetal cells in a mother may increase the risk of autoimmune disease and play a role in cancer, although the details remains unclear. The effects of maternal cells in offspring are even less clear, but there is little doubt that they have the potential to carry biological memory from mother to child.

Other forms of biological memory may result from chemical changes to the cellular protein/DNA complexes known as chromatin. These alterations lead to persistent changes to cells that are independent of the actual sequences of genetic information encoded in DNA.

Even though these epigenetic effects do not change the sequence of DNA encoding information, they can nonetheless be inherited (across cell divisions and potentially across generations). In many instances, such epigenetic modifications are no less profound than genetic mutations; they can alter the way that your DNA sequence is interpreted and expressed and change the way that genes are turned on and off in specific tissues in response to environmental cues. This control of gene expression varies strongly across different tissues and different times in development. Epigenetic changes thus act like marginal notes and highlighting, influencing the interpretation of the genetic instruction manual for making a human being.

One of the major mechanisms of epigenetic marking is the addition of a methyl group (one carbon atom bonded to three hydrogens) to specific sites within DNA molecules. The addition of this methyl group changes the shape of the DNA and alters how it interacts with the machinery that reads gene sequences and controls their expression. Similarly, methyl groups or acetyl groups (two carbons, three hydrogens, and an oxygen) can be added to specific amino acids within the DNA chromatin complex, making it easier or more difficult to read the underlying DNA sequences. This influences gene expression and, thus, the structure and function of cells. Environmental factors can alter DNA methylation patterns, thereby affecting the way that your DNA sequence is interpreted. For example, the DNA methylation patterns of people who were in utero when their mothers experienced the Dutch Hongerwinter are different from the methylation patterns of people whose mothers did not experience famine during their early development.

Epigenetic alterations of DNA can be passed from one cell to its descendants, so environmental factors that affect these epigenetic signatures can persist throughout life. In plants and in some animals, changes in DNA methylation can even be passed across generations (up to forty generations in some cases). This type of multigenerational transmission is less likely in mammals (including humans) whose epigenomes undergo substantial reprogramming during early development. But at least some of the genome in mammals is now thought to be resistant to this reprogramming, so the number of generations over which biological memory can be conveyed in humans and other mammals remains an open question.

Although DNA methylation patterns are thought to be largely reset in humans during early development, other mechanisms of epigenetic inheritance have the potential to be passed across multiple generations. In humans and other mammals, the fertilized egg that gives rise to a human being was initially formed in a mother's ovaries when she is an eight-week-old embryo during her own mother's pregnancy. Thus, the fertilized egg each of us came from carries information about the environment during our grandmother's pregnancy that has the potential to influence the pathways and patterns of gene expression during our lifetimes, and similar effects hold for formation of sperm in males. In other words, the environment experienced by both your grandmother and grandfather may have effects on you.

The environment you experience in the first few years of life can also have long-lasting effects on physiology, metabolism, stress responses, and emotions in later life. For example, in rats, offspring of neglectful mothers (who feed but do not lick and groom their babies as much as more nurturing mothers) tend to grow up to be anxious, while highly nurtured pups tend to grow up to be calm. These quantifiable behavioural effects are correlated with changes in the epigenome. Similar patterns occur in stickleback fish. In this species, it is the father that provides parental care, fanning the eggs in the nest and looking after the offspring. Offspring of neglectful stickleback fathers tend to be more anxious than offspring of more nurturing fathers. Interestingly, these effects are different in male and female offspring, and studies in mammals show similar sex-specific effects. Epigenetic observations in humans, such as those of the Dutch Hongerwinter survivors, suggest that the same may be true in humans and that the effects of maternal, paternal, or early-life stress may differ between the sexes.

The more we learn about biological memory through the study of epigenetics, the more we realize how profoundly our past experiences, and even those of our parents and grandparents (and maybe beyond), can influence our lives. Our present-day environment also affects our epigenome, and we are thus influenced by a complex interweaving of our current circumstances and choices, our past experiences, and those of the generations who came before us.

# RECONCILIATION POLE

*Nicola Levell*

THE AMERICAN ART CRITIC and philosopher Arthur Danto asserted, "We erect monuments so that we shall always remember, and build memorials so that we shall never forget."[1] Totem poles – archetypal symbols of Indigenous Northwest Coast culture – function as both monuments and memorials. They are bearers of oral histories and genealogies as well as commemorative markers of past events and deceased individuals. Raised in 2017 and standing seventeen metres high on the Main Mall of the Vancouver campus of the University of British Columbia (the unceded territory of the Musqueam First Nation), *Reconciliation Pole* is a memorial that tells the horrific history of Canada's Indian residential schools and charges us never to forget as we move towards the future and reconciliation.

Canada's Indian residential schools were in operation for over a century, with the last one closing in 1996. Their goal was assimilation of Indigenous children into the body politic: they were state-funded, often church-run institutions of physical violence and emotional and sexual abuse. Government documents estimate that at least 150,000 First Nation, Metis, and Inuit children passed through the residential school system. On arrival, they were generally given a haircut, a uniform, an Anglicized name, and an identification number, and they were prohibited from speaking Indigenous languages. With callous disregard for human rights, Indigenous children were torn from their families and stripped of their social skin, identity, and culture; they were denied access to their community frameworks and practices of collective memory. The multigenerational effects of this loss and trauma are still starkly visible, lived, and felt.

How can memorials, such as *Reconciliation Pole*, play a role in the process of healing and acknowledging Indigenous peoples' historical experience, memory, and loss? How can memorials educate, inspire hope, and reshape public memory, as Canada's Truth and Reconciliation Commission, or the TRC, anticipates? How can collective remembrance of a shameful past be transformed into artistic expression, affirmative action, and constructive dialogue as we chart a course for reconciliation? To address these questions, it is insightful to explore not only the material and visual symbolism of *Reconciliation Pole* but also the collective-memory practices and protocols of the ceremony that marked its installation on the UBC campus. Monumental outdoor artworks, including memorials, are often overlooked and passed by; they depend on public engagement, commemorations, rituals, and ceremonies to bring them to life.

The idea of memorials as potent and affective sites of collective remembrance, healing, and reconciliation is detailed in Volume 6 of *The Final Report of the Truth and Reconciliation Commission of*

*Canada*: "True reconciliation can take place only through a reshaping of a shared, national, collective memory of who we are and what has come before ... As Canadians gather in public spaces to share their memories, beliefs, and ideas about the past with others, our collective understanding of the present and future is formed. Public memory is dynamic – it changes over time as new understandings, dialogues, artistic expressions, and commemorations emerge."

On April 1, 2017, thousands of people gathered on UBC's Vancouver campus to participate in the ceremony "Reconciliation Pole – Honouring a Time Before, During, and After Canada's Indian Residential Schools." This public event, which unfolded over the course of five hours, was envisaged as an opportunity for Indigenous and non-Indigenous participants to come together to witness and to remember the injustices of the past and the present, to raise the pole, and to foster collective memory and intercultural understanding, as part of our shared commitment to reconciliation. The catalyst for this ceremony of public remembrance and healing was the extraordinary *Reconciliation Pole*. Carved on Haida Gwaii from an eight-hundred-year-old *ts'uu* (red cedar) over a two-year period (2015–17), this unique pole was designed and executed by 7idansuu, James Hart, a Haida master carver and hereditary chief of the Saanggalth Stastas Eagle Clan, along with his team of carvers: Gwaliga Hart, John Brent Bennett, Brandon Brown, Jaalen Edenshaw, Leon Ridley, Derek White, and his late son, Carl Hart. Like other Haida monumental poles, or *gyáa'aang*, *Reconciliation Pole* – said to be the largest, if not the tallest totem pole in existence – is a narrative structure. In this case, it is intended to be "read" from the bottom up, and the

## WHAT STORY DOES RECONCILIATION POLE TELL?

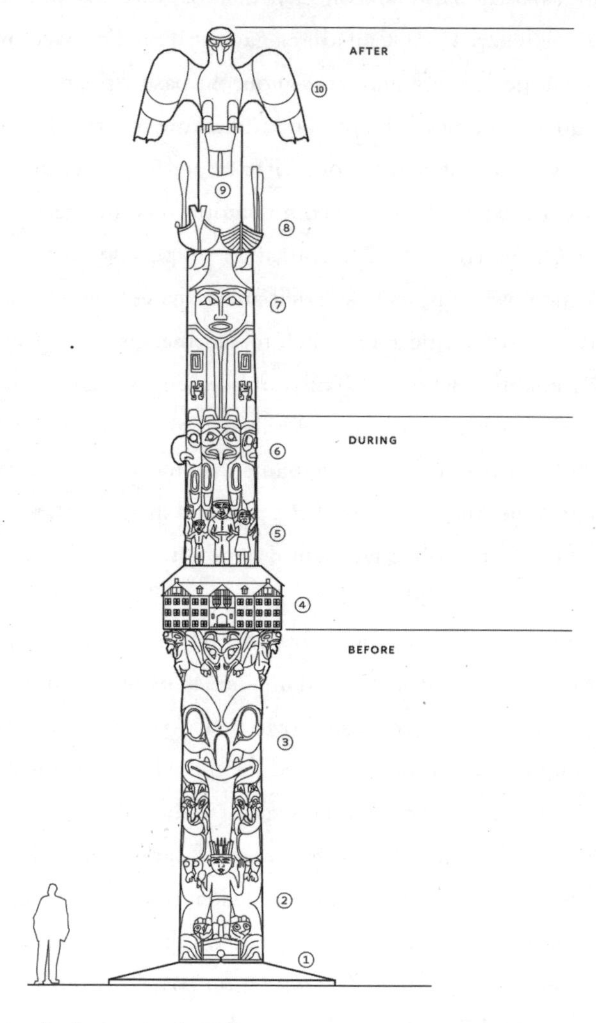


*Reconciliation Pole* diagram. Photo by DNA Engineering. Courtesy of *UBC News*.

Haida poles are read from bottom to top.

1) Surrounding the base of the pole are salmon symbolizing life and its cycles.
2) Between the legs of Bear Mother is sGaaga (Shaman), who stands on top of Salmon House and enacts a ritual to ensure their return.
3) Bear Mother holds her two cubs, Raven looks out from between Bear Mother's Ears.
4) A Canadian Indian residential school house, a government-instituted system designed to assimilate and destroy all Indigenous cultures across Canada.
5) The children holding and supporting one another are wearing their school uniforms and numbers by which each child was identified. Their feet are not depicted, as they were not grounded during those times.
6) Four Spirit Figures: killer whale – water, bear – land, eagle – air, Thunderbird – the supernatural. They symbolize the ancestries, environment, worldly realms, and the cultures in which they are rooted, that each child came from.
7) The mother, father, and their children symbolize the family unit and are dressed in traditional high-ranking attire symbolizing revitalization and strength of today.
8) Above the family is the canoe and longboat shown travelling forward, side by side. The canoe represents the First Nations and governances across Canada. The longboat represents Canada's governances and Canadian people. This symbolism respectfully honours differences, but most importantly displays us travelling forward together side by side.
9) Four Coppers, coloured to represent the peoples of the world, symbolize and celebrate cultural diversity.
10) Eagle represents power, togetherness, determination, and speaks to a sustainable direction forward.

The copper nails covering areas of the pole are in remembrance of the many children who died at Canada's Indian Residential Schools – each nail commemorates one child.

visual narrative is organized into three sections: "Before," "During," and "After" the Indian residential schools.

The representation of the residential school in the central section of the pole is based on the Coqualeetza Industrial School. The school opened in 1886 near Chilliwack, British Columbia, and 7idansuu, James Hart's relatives, including his grandfather, great-aunts, and uncles, were sent there. In *Reconciliation Pole*, children stand on top of the school, linking arms and holding on to one another for support; in some cases, they wear uniforms and are marked with identification numbers. Their feet are not represented, indicating that they were not grounded during this time. In a conscious move to incorporate other Indigenous peoples into the narrative and the aesthetics of the pole, 7idansuu, James Hart invited ten Indigenous artists, from Canada and the United States, to carve and paint the faces of the children and embellish their figures.[2] One child's face is left uncarved – smooth, ghost-like, without features – to denote and remember all the unknown children who suffered in Indian residential schools.

In addition to the visual narrative, the materiality of the pole is richly infused with cultural meanings and symbolism, from its red cedar body, through the abalone-inlaid headdress, to the four copper shields at the top, which are coloured to symbolize Canada's cultural diversity. But the most poignant symbols of all are the thousands of copper nails, precious beings that are clustered on *Reconciliation Pole*: each one commemorates a child who died while at residential school. On the underside of the residential school, copper nailheads are shaped into two haunting, stylized skeletons, symbolizing the children's bodies buried in the cemeteries of these negligent institutions. The TRC's official records reveal that more than six thousand children died at residential schools, while other sources suggest tens of thousands more perished in them through maltreatment and neglect. In an act of collective remembrance, residential school survivors, members of their families, and many other Indigenous and non-Indigenous participants hammered the copper-bodied nails into *Reconciliation Pole*.

Survivors of the Indian residential school system spoke at the *Reconciliation Pole* ceremony. Their speeches were suffused with painful childhood memories of cruelty and abuse. At the same time, the speakers conveyed a sense of optimism, which reflected the vitality, resilience, and strength of these First Nations Elders and their cultures. In accordance with First Nations protocol, the ceremony began with a Musqueam welcome, enunciated in hən̓q̓əmin̓əm̓ and English by sʔəyəɬəq Elder Larry Grant, who welcomed Indigenous and non-Indigenous guests to the Musqueam Nation's unceded ancestral lands. This welcoming was followed by official speeches by Musqueam and Haida chiefs and councillors, by UBC president Santa Ono, and by the Indian residential school survivors. Linc Kesler, director of UBC's First Nations House of Learning and senior adviser to the president on Aboriginal affairs, acted as master of ceremonies. Throughout the afternoon, there was mindful attention to Indigenous protocols and diplomacy in speeches and actions: welcoming, honouring, songs, dancing, gifting. Highly visible was ceremonial regalia: woven blankets and hats, inlaid headdresses and ornate accessories, painted rattles and drums. These forms of regalia are also represented on *Reconciliation Pole*. In the upper section, "After," the family group is depicted wearing regalia, which symbolizes the revitalization of Indigenous cultural heritage, practices, and memory.

To historically contextualize the structure of the commemorative ceremony and understand its significance for reshaping collective memory and reconciliation, it is important to acknowledge that the Musqueam and other Salish First Nations do not have a tradition of totem pole carving. For many decades, it was insulting and hurtful for them to witness the installation of these memorial and territorial markers, belonging to northern groups such as the Haida, on their unceded territories. Institutions of the colonial past and recent present failed to acknowledge Musqueam title and rights to their ancestral land: Musqueam people were not consulted on what was placed on this land, done with it, or extracted from it. At the *Reconciliation Pole* ceremony, repeated reference was made to the Indigenous protocols of hosting, nonadversarial understanding, and mutual respect. Elder Larry Grant and Wayne Sparrow, elected chief of the Musqueam Indian Band, both acknowledged and thanked the Haida Nation, UBC, and everyone present for following proper protocol.

Another reference to Indigenous protocol, rooted in First Nations frameworks and the practices of community memory, was the call for us to bear witness. The act of bearing witness, which was articulated in official speeches and in the printed program of the ceremony, is identified in the directives of the TRC as a fundamental "Aboriginal principle" to be adopted as we move towards reconciliation and forge new relationships. Unquestionably, witnessing is a purposeful act that contributes to the formation of collective memory. As witnesses, we are charged to actively remember, to assume an ethical responsibility, to publicly recall and transmit the knowledge of that which we are invited to observe. Although the TRC's cooption of witnessing as a pan-Indigenous principle has been critiqued by scholars, its employment in the case of *Reconciliation Pole* was culturally appropriate because witnessing is an established, historical, and contemporary dimension of Indigenous ceremonial practices on the Northwest Coast, including pole raisings.

While the act of witnessing respects Indigenous memory practices and teachings, it is important to recognize that the tenets of reconciliation are contested by some Indigenous scholars and communities. The Metis academic, curator, and artist David Garneau argues that the idea of reconciliation, in the Canadian context, is deeply flawed because reconciliation means repairing relationships that were once amicable and harmonious. He also points out that the TRC's project of reconciliation is based on non-Indigenous practices of healing and closure, which are focused on the individual testimonies of victims rather than those of perpetrators. He directly relates this "confessional" mode of disclosure to a colonial, Catholic ideology. In moving forward, he recommends that we reframe the dialogue, between Indigenous and non-Indigenous people, as a process of conciliation. Conciliation is an alternative dispute-resolution process that brings parties together to work through problems and differences and find mutually acceptable solutions. So, Garneau asks, "How are we to change the sites of Reconciliation into sites of Conciliation? How do we prevent Reconciliation from being primarily a spectacle of individual pain for settler consumption and Aboriginal shame?"[3]

*Reconciliation Pole* played and will play a part in this collective process of reshaping public memory and conciliation. The raising of the pole had impacts that were felt – through emotion and sensory experience, through engagement and empathy – between and among Indigenous and non-Indigenous participants. The ceremony constituted an act of coming together in the same space to witness and remember; to listen to the speeches; to observe the blessing of the pole and the carver's dance; to pull the ropes and raise the pole; to watch the dancing and feast together. These activities were embraced with the possibility of moving forward together. This future course was captured in the words of Adina Williams, a UBC student from the Squamish Nation who was the last speaker before the pole raising. The daughter of a residential school survivor, grounded and calm, she spoke: "My parents, as survivors, have used that strength and resilience to help me become the proud Squamish and 'Na̱mgi̱s person I am today. Before I left the house this morning, my parents reminded me that just me standing here today, and being able to speak to you and share how I think we can move forward together towards a reconciled future – that is an opportunity that my parents did not have. It is gatherings like these that leave me very hopeful for a future for my children, grandchildren and great-grandchildren."

This hope for the future is expressed on the top of *Reconciliation Pole* in the form of the majestic eagle about to take flight. It represents, to recall 7idansuu James Hart's words: "The power and determination needed to look towards the future."[4] He further contends, "We still need to move forward ... Canada needs to stand up ... not just apologizing, but really acknowledging what happened in the past, so it doesn't ever happen again."[5] In moving forward, it is vital that *Reconciliation Pole* continues to be engaged with as an artistic expression and as a site of remembrance, learning, and dialogue. In this way, this unique memorial will help shape collective memory and mutual understanding; it will help ensure, as Danto explains, that we never forget the losses suffered, the victims, and the horrors of the past.

ACKNOWLEDGMENTS

I thank 7idansuu James Hart for giving so freely of his time and knowledge to talk about the history and making of *Reconciliation Pole*. I also extend my gratitude and thanks to the following individuals for their help, insights, or permissions: Carole Blackburn, Jisgang Nika Collison, Leslie Dickson, Gwaliga Hart, Linc Kesler, Naomi Sawada, Patricia Shaw, Christopher Smith, Ginaawaan Darin Swanson, Jana Tyner, and Adina Williams. *Reconciliation Pole* was commissioned by the Audain Foundation and UBC.

NOTES

1. Arthur C. Danto, "The Vietnam Veterans Memorial," *The Nation*, August 31, 1985, 152.
2. These artists are Corey Bulpitt (Haida), Kevin Cranmer (Kwakwa̱ka̱'wakw), Reg Davidson (Haida), Phil Gray (Cree), Sven Haakanson (Aleut), Greg Hill (Mohawk), Zacharias Kunuk (Inuit), Shane Perley-Dutcher (Maliseet), Susan Point (Musqueam), and Christian White (Haida).
3. David Garneau, "Imaginary Spaces of Conciliation and Reconciliation: Art, Curation, and Healing," in *Arts of Engagement: Taking Aesthetic Action in and beyond the Truth and Reconciliation Commission of Canada*, ed. Keith Martin, Dylan Robinson, and David Garneau (Waterloo: Wilfrid Laurier University Press, 2016), 29.

[4] Leslie Dickson, “Reconciliation Pole Installed on UBC Vancouver Campus,” *UBC News*, March 30, 2017, https://news.ubc.ca/2017/03/30/reconciliation-pole-to-be-installed-on-ubc-vancouver-campus/.

[5] Hadani Ditmars, “A B.C. Reconciliation Totem’s Lessons on How to Move Forward,” *Maclean’s*, June 30, 2017, http://www.macleans.ca/news/canada/a-b-c-reconciliation-totems-lessons-on-how-to-move-forward/.

# FIRST LIGHT

*Harvey Richer*

CULTURAL HISTORIANS examine evidence of past civilizations over thousands of years, while anthropologists and archaeologists record the history of humanity back to the origin of our species. Evolutionary biologists trace the history of life back to the early days of our planet, while geologists study the earliest chemical and biological processes on Earth through signatures left in ancient rocks and ocean basins. But how are we, tied to the confines of Earth, to explore the distant memory of our universe, which goes back billions of years, when our planet was no more than a fleeting cloud of dust scattered across the widest reaches of space? What are we to make of the events that occurred well before our own planet came into existence?

The universe itself provides a physical memory of past events, recorded in cosmic radiation that can take on various forms, including cosmic rays, light, and gravitational waves. The speed of light, almost 300,000 kilometres per second, is almost unimaginably fast. On Earth, we experience this as an instantaneous perception of light emitted from all objects. Over the vast reaches of the universe, however, light travel can take a long time, often millions or even billions of years. Our nearest big galactic neighbour, the Andromeda Galaxy, is located 2 million light years from us, meaning that light from this galaxy takes 2 million years to reach an observer on Earth. The light from a supernova exploding in the galaxy will take about the same amount of time to become visible to us. On the other hand, the light from the most distant known galaxies takes more than five thousand times longer (10 billion years) to arrive at Earth's surface. Like a faint memory echoing across innumerable generations, information about major events in our universe (our universal memory) is transmitted to us with a time delay much longer than the entire history of humanity. We see distant galaxies as they were in their youth, shortly after the universe was created through a massive explosion known as the Big Bang.

Astrophysicists explore the history of the universe by peering into the dark night sky, using powerful telescopes to search for nearly imperceptible spots of light emanating from distant galaxies. While 2018 marks the centennial of the end of the First World War, it also marks the one hundredth anniversary of the opening of the large 1.8-metre reflecting telescope in Victoria, British Columbia. As the first big science project for Canada, the telescope, which has more than fifty thousand times the light-gathering power of the human eye, was to be the largest in the world. Construction was initially approved in 1910, but delays, largely due to a scratch on the mirror that forced it to be reground, caused its "first light" to come six months after the 2.5-metre Hooker Telescope became operational at the Mount Wilson Observatory in California. The mirror for the Victoria telescope was delivered from the

Bringing supplies up Little Saanich Mountain, Victoria, to construct the 1.8-metre telescope, likely in 1916. The nearly completed telescope dome is seen in the background. Courtesy National Research Council of Canada.

manufacturers in Belgium only a week before the outbreak of the war.

Although the end of the war and the opening of what was supposed to be the world's largest telescope in Canada are unconnected, both events shaped the world in their own profound ways. Early observations with the Victoria telescope of the spectrum of interstellar gas in our galaxy uncovered a background radiation field with a temperature of about two degrees above absolute zero (i.e., −271°C). Unbeknownst to those early observers, their work had uncovered a faint echo of the Big Bang event that had initiated the expansion of the universe. Over time, the presence of this ultralow temperature radiation field has come to represent the most compelling line of evidence in support of the Big Bang theory (the idea, not the television program). In the early twentieth century, when the telescope was first in operation and the world was ripped apart by the ravages of the First World War, there was no theoretical foundation that could have connected these early observations to a proper cosmological interpretation. As a result, the data languished, only to be correctly interpreted decades later with the benefit of many additional observations and new theoretical models. This later work was recognized in 1978 with a Nobel Prize in physics, awarded to a team of scientists, none of whom was Canadian. Meanwhile, California's Hooker Telescope hardly sat idle during the early years after its commissioning. Its large mirror, coupled with a somewhat better location than Victoria, allowed Edwin Hubble to discover that the universe was expanding and that the most distant galaxies were moving away from one another at the highest speeds. This expansion represents the cosmological memory of the birth of the universe.

The presence of the 1.8-metre telescope in Victoria did much to spur the development of Canadian astronomy in subsequent decades. Generations of Canadian astronomers used either the telescope itself or its data to study a wide body of phenomena related to solar system objects, stars, and galaxies. This work put Canadians at the forefront of international astronomical research, providing an impetus for additional Canadian funding of more recent frontier telescopes. In 1979, the 3.6-metre Canada–France–Hawaii Telescope became operational.

Among many other achievements, astronomers have used this telescope to image very distant supernovae, helping to show that the expansion of the universe is accelerating. Two decades later, Canada joined in a partnership of twin 8.2-metre telescopes (one in each of the southern and northern hemispheres). These telescopes, together with other instruments, were used by a Canadian astronomer to provide the first images of a multiplanet system outside our own solar system.

In recent years, Canada has joined in a consortium with China, India, Japan, and the United States to construct a massive 30-metre telescope, called TMT, on the Big Island of Hawaii. This telescope, when completed, will be one of the largest in the world, enabling astronomers to answer some of the most fundamental questions in science such as the composition of the universe, when the first galaxies formed, and whether there is life elsewhere in the universe.

But the TMT project is evolving in ways not expected when the project began nearly two decades ago. The proposed location on Mauna Kea in Hawaii placed the telescope in tension with a general resurgence in Hawaiian identity, language, and culture, leading to vigorous protests against the project. When the TMT project began in 2000, only about 0.1 percent of Hawaiians still spoke their native language. This loss of language reflected a history of colonialism and a large-scale erasure of cultural memory, which has also suppressed art, dance, and other forms of Indigenous knowledge. Seen through this lens, the controversy surrounding the telescope project can be viewed as reflecting a tension between cultural and astronomical memory.

There are other cosmic images that constitute a universal memory and advance understanding of our (and potentially other) life in the universe. In 1977, the Voyager 1 spacecraft was launched. Its mission was to carry out a grand tour of the solar system, making close passages of all the planets except Pluto, which was still at that time considered a planet (it was demoted to a minor planet in 2006). Voyager 1 meticulously photographed each planet and many of their moons, producing images that are still the best existing representations of these extraterrestrial systems. In 1990, when Voyager 1 was 6.4 billion kilometres away from Earth, the eminent astrophysicist Carl Sagan (at that time a key member of the Voyager mission imaging team) requested that the spacecraft be turned around so that it could photograph Earth from this enormous distance. NASA was not pleased with the request, but Sagan prevailed. In the resulting image, Earth appears as a tiny blue dot, less than a single pixel wide, suspended on a beam of sunlight scattered in the camera optics. This image remains one of the most iconic photographs of the entire space program. It records, as no other image does, our incredibly small place in the vastness of space. The only other image of comparable impact may be that of the first footprints left by astronauts when they landed on the Moon in 1969. Because of the almost complete lack of erosion on the lunar surface, those footprints could conceivably last as long as the Moon does (about 5 billion years more), archiving the memory of the first human voyage beyond Earth.

One of the most important areas of current research in astronomy (and perhaps in all of science) is the search for planets beyond our own solar system, particularly those that might support life. Several decades ago, there were no known planetary systems orbiting around any star other than our own. Today, we are confident that there are more planets in our galaxy than there are stars. Of particular interest is the search for ancient solar systems. If such systems can be located, they could potentially signal an earlier rise for life in the universe, well beyond the approximately 4 billion years that life has existed on Earth.

Ancient star clusters are one potential host for early life-supporting systems. We can accurately determine the age of such systems, which often house up to a million suns. Because planets are generally seen only by reflected light from their host star, they are often a 100 million or more times fainter than the stars they orbit, making it extremely difficult to isolate a very old planetary system in which the planets are still intact and circling their star. We can, however, see evidence of planets that have been destroyed by their host stars. This happens when a star runs out of nuclear fuel, expands enormously, and then contracts to form a white dwarf star, an object about the size of Earth and about half the mass of the Sun. The dusty remnant material from the doomed planet and its moons and comets remains in orbit around the white dwarf in a disc shape, and it is this material that telescopes may detect as a dim infrared glow. This is, of course, not necessarily a signature of life, but it does provide evidence that a planetary system once existed around that star – surely a necessary condition for life as we know it.

Interestingly, the third white dwarf star discovered, just a few months before the end of the First World War, turned out to be just such an object. It exhibits the presence of elements such as carbon, nitrogen, and calcium that are heavier than hydrogen and helium and are thus expected to rapidly sink below the atmosphere in such a dense star, effectively becoming invisible to outside observers. The persistent signature of these heavier elements thus implies a source of constant replenishment from the destroyed planets orbiting in a disc around the star, a clue to potential extraterrestrial conditions that might conceivably support life.

Memory, the ability to store and recall information, has traditionally been attributed to organisms or populations. But over the last century the rise of sophisticated electronics and computing systems has introduced new notions of nonliving synthetic memory associated with inanimate substances, such as memory chips and other information-storage devices. It may be that the most profound memory is that of the universe itself. Astronomers peering into the night sky have long sought to decipher that memory. In a quest for deep understanding, we may be on the verge of tapping into signatures of the very first life harboured in the universe and the ultimate origins of our existence.

# CORROBORATION

*Elee Kraljii Gardiner*

I try to line it up neat as the rows in the movie theatre
but each passing moment further muddles memory.
We were eating popcorn and then witnessing an execution
three rows behind the murder victim. The shooter
ran out before we knew what happened. How do we make
sense of that. I plastered myself to the Sprite-sticky floor.
We swam among people whose clothes were ruined
with the victim's blood and bone. I remember thinking, "Why
rush to the streets where the killer might be?" It seemed important
to go far away from the screaming girlfriend.
What if he came back. What if retaliation. Would we
bump into the murderer outside. We stayed near the entrance
with an impulse to walk away but leashed to the idea of responsible
witnessing. Did someone call the cops. *I'll call them.* Do we need
to stay. *Let's go*, but we couldn't. An ambulance charged forward.
I remember thinking, "What's the rush? He's dead." We walked away

tentatively, like we had done something wrong. At the car, I touched
the door handle but didn't pull. Later, the police came into our living
room for witness statements. Past midnight. Their dull black guns.
Boots the size of houses. A bent pen. Our details down.
We had different answers about his clothes, when he looked
over his shoulder. We remembered the movie the same, though.
What I was wearing. Later, we learned names for the two enemies,
but they escape me. I remember where we sat. I check exists now.
I never forget to do that.

# SHIPS AT SEA

*Renisa Mawani*

As if it were too great, too mighty for common virtues, the ocean has no compassion, no faith, no law, no memory.

– Joseph Conrad, *Mirror of the Sea*

Where are your monuments, your battles, martyrs?
Where is your tribal memory?

Sirs, in that grey vault. The sea. The sea has locked them up. The sea is History.

– Derek Walcott, "The Sea Is History"

THE EVENTS WE REMEMBER and forget and the histories we tell are inextricably shaped by land. Migration histories, to draw but one example, are most often recalled through arrival and departure.

Although seaborne travel persisted as the primary means of passenger transport until the mid-twentieth century, long ocean voyages rarely figure in narratives of migration. How might forced and so-called free passages be recollected and retold when the sea is repositioned as a force of history and a site of memory?

On July 18, 1914, the *Literary Digest*, a leading New York magazine, published a commentary titled "Sikhs Besieging Canada." The first line read, "The lappings of the Asiatic tide upon our western shores remind some writers that almost every country on earth has at some time been overrun by a great racial migration, and that even our own presence here is the result of such a movement." The focus of the article then shifted from "every country on earth" to Canada and the SS *Komagata Maru*, a British-built and Japanese-owned steamship that dropped anchor in Vancouver Harbour on May 23, 1914. As the *Literary Digest* circulated among its elite and mostly white readership, all but 20 of the 376 Punjabi travellers – under the leadership of Baba Gurdit Singh – were refused landing. For two months, the passengers were detained aboard the vessel under deplorable conditions. On July 23, 1914, the ship was deported from Vancouver to Calcutta. Four days into its transpacific journey, the First World War commenced. For most of the passengers, the Great War would carry serious consequences, including imprisonment and, for some, death.

References to oceanic tides and waves may not be common today but, historically, they shaped and even animated discussions of migration. In the early twentieth century, as the *Literary Digest* makes clear, colonial administrators often drew on ocean metaphors to underscore the dangers of Chinese, Japanese, and Indian migration to the white settler Dominions. Oceans offered evocative imagery through which authorities expressed their racial fears of seaborne travel from East to West. But what the *Literary Digest* left unsaid is that oceans were equally important as material sites of death and of hope. For those above and below deck – including slaves, indentured labourers, and migrants, including the *Komagata Maru* passengers – the ocean was vital to their survival. It encompassed and embraced the vessel, determined the perilous conditions of travel, and extended into an unknown future, offering new horizons of possibility. For Gurdit Singh, who chartered the *Komagata Maru*, oceans were spaces of colonial violence but ones that inspired alternative imaginaries of freedom. The sea, as Singh envisaged it, generated new and renewed opportunities for anticolonial solidarities and struggles against British imperial control.

In the European imagination, oceans have featured prominently in narratives of adventure, conquest, commerce, and capital but not often as spaces of history. For some European philosophers, Europe was deeply enmeshed with the sea, whereas non-Europe was associated with land.[1] Despite these associations, however, the sea was thought to be an empty space, one without history, politics, or law. Joseph Conrad, a seafarer turned literary virtuoso, was intimately aware of the significance of oceans both to human experience and imagination. Conrad drew his literary inspirations from his many years of seafaring and devoted much of his creative attention to portraying life in maritime worlds. Yet in his literary works, as the opening epigraph makes clear, oceans are rarely perceived as having a past or a history: "The ocean has no compassion, no faith, no law, no memory."[2]

What, then, is required to understand the sea as a critical site of collective memory? The answer to this question demands a geographical, historical, and ethical reorientation, one that directs our attention away from terra firma to the vast, interconnected, and aqueous regions of the planet. At the very outset, this elemental shift dramatically alters what we think of as the past, of remembering and forgetting, of memory and history. Historical evidence is no longer composed only of texts, files, or documents. It also includes stories and memories, artifacts and objects – bodies, fragments, and shipwrecks – that continue to animate life and death. The sea as memory invites a geopolitical journey from Europe, as the presumed historical origin and cultural centre of the earth, to other regions and their attendant histories. It moves us from two-dimensional surfaces of the sea into tumultuous worlds of vertical depths. The sea as memory shifts understandings of history away from western Europe's great wars and epic battles, including the First World War, to subaltern and racial histories that have been central to the making of the modern world – including the forced movements of slaves, indentured labourers, and migrants – histories that remain submerged, both figuratively and, in some cases, literally at the bottom of the sea. As Derek Walcott, the St. Lucian poet insists, it is only when we view the sea as history that we can begin to recuperate these lost, forgotten, and unspeakable events that have been erased from historical memory.[3] Though Walcott's point of reference is transatlantic slavery, the sea as memory opens other narratives of racial violence and of hope that continue to shape migration.

Histories of the *Komagata Maru* have typically been written through landfall and territoriality. While these accounts are crucial to

Canada's national memory, they have obscured other equally important narratives, including those of the ship at sea. The *Komagata Maru* crossed the Pacific, first from Hong Kong to Vancouver and then from Vancouver to Yokohama. It also voyaged across the Indian Ocean on its journey to Calcutta. Most of the passengers were confined to the ship for six long months, without ever setting foot on land. The vessel's restricted space, approximately one hundred square metres, was thus a stage upon which the lives of more than 350 people unfolded for half a year. Yet we know little of the events that occurred at sea. How, for example, did Sikh, Muslim, and Hindu passengers interact, maintain collective hope, and support one another? In this case, as in many other ocean crossings, what happened between departure and arrival is not a focus of analysis; the ship, storms, winds, and salt water are forgotten. They never enter as forces that shape memory or make history.

The ocean has long been a subject of archival production. From the sixteenth century onwards, the movement of European ships, passengers, and cargos demanded copious amounts of documentation. Ship registries, slave manifests, captains' ledgers, maps, letters, and diaries have informed literature and art and have provided the sources we use to write history. On the vast and open sea, these texts operated both as records and as methods of regulation. When merchant ships entered port, shipmasters were required to submit detailed manifests of their passengers and cargos. Captains, crews, and travellers spent long and arduous months at sea. Reading and writing became ways of passing time. Shipping records have become raw materials for composing the past. Yet the ocean itself continues to be seen as a blank space, a space beyond history.

In many ways, this view of the sea as empty reflects a perspective from the North Atlantic. For Caribbean writers and scholars of the Indian Ocean, by contrast, the sea is alive – filled with moving boats, social and religious networks, kinship relations, and the spirits of ancestors, creating and re-creating routes from the past. Viewed in this light, the ocean is a rich treasure trove of overlapping, intersecting, and competing memories that may not necessarily be written or recorded. Their traces reside in stories of journeys and places and also in fragments, some dwelling at the bottom of the sea. Whereas Walcott views the sea as history, British Guyanese poet Fred D'Aguiar describes the sea as slavery. The two are inextricably linked. The sea directs us to histories of forced and free migration: European resettlement, transatlantic slavery, and forms of racial subjection, deracination, and death. The sea as history and the sea as slavery are not buried in the past. Rather, these histories continue to recur in perilous seaborne journeys, as the migrant boats in the contemporary Mediterranean remind us.

The sea as memory demands that we look beyond the written texts of history to other vestiges of the past. One of these may be the ship itself. The vessel that became the *Komagata Maru* spent thirty-six years at sea. It was born as the *Stubbenhuk* (1890–94), then the *Sicilia* (1894–1913), then the *Komagata Maru* (1913–24), and finally the *Heian Maru* (1924–26). In its first two incarnations, the vessel transported European settlers from German and, later, Mediterranean cities to ports of call along the eastern seaboard, to Indigenous lands now known as Canada and the United States. The ship anchored at many of the same ports as the slave ships that preceded it: Boston,

Baltimore, New Orleans, Halifax, and Ellis Island. Renamed the *Komagata Maru*, the vessel transported coal across the South China Sea and was scheduled to enter the Chinese indenture trade before the owners chartered it to Gurdit Singh. In 1926, the *Heian Maru* was shipwrecked off the coast of Hokkaido. Today, it sits at the bottom of the sea, where the North Pacific meets the Sea of Japan.

Oceans comprise 70 percent of Earth's surface and have dramatically shaped the experiences of movement and mobility, including the forced passages of slaves, migrants, and refugees and the transport of persons and commodities required to sustain global capitalism.[4] Conceptualizing the sea as memory centres the planet's watery regions, directing our attention to seaborne histories, especially those of racial violence and anticolonial struggle. In the case of the *Stubbenhuk*, *Sicilia*, and *Komagata Maru*, when the ship is read as a historical artifact, it gestures to the intersecting and overlapping histories of Indigenous dispossession, transatlantic slavery, Chinese indenture, and so-called free migration, or what colonial authorities referred to as "the Asiatic tide." But the sea as memory also points us to other possibilities for remembering the past and imagining the future. Like the sea, history and memory cannot be easily defined either as objective or subjective. They are aqueous forces – written and unwritten, visible and invisible, solid and fluid – that join distant geographies and other times. These histories are not past; they are in the present, in the now, in the sea.

The sea as memory dramatically challenges our concepts of time and space, history, violence, and freedom. Though Conrad and Walcott saw the ocean in distinct ways, from divergent hemispheres and through specific histories, for both writers, the world looked radically different from the sea. The sea as memory invites us to remember what we cannot see, what is forgotten but always there: a past that has shaped and informed our global condition, a past hovering at the edges of a future unfolding, a past we must remember for the future to be otherwise.

NOTES

1 I refer to the work of Georg Hegel and Carl Schmitt. On the former, see Alexis Wick, *The Red Sea: In Search of Lost Space* (Berkeley: University of California Press, 2016), 57–65. See also Carl Schmitt, "The Planetary Tension between Orient and Occident and the Opposition between Land and Sea," *Politica Comun* 5 (2014): 1–19, http://dx.doi.org/10.3998/pc.12322227.0005.011.

2 Joseph Conrad, *Mirror of the Sea* (London: Meuthen and Co., 1906).

3 Derek Walcott, "The Sea Is History," in *Derek Walcott: Collected Poems, 1948–1984* (New York: Farrar, Strauss, and Giroux, 1986), 364–67. The ocean has figured prominently in histories of slavery. See Fred D'Aguiar, *Feeding the Ghosts: A Novel* (New York: Harper Collins, 1997); Saidiya Hartman, *Scenes of Subjection: Terror, Slavery, and Self-Making in Nineteenth-Century America* (Oxford: Oxford University Press, 1997); M. NourbeSe Philip, *Zong!* (Middletown, CT: Wesleyan University Press, 2008); Christina Sharpe, *In the Wake: On Blackness and Being* (Durham: Duke University Press, 2016); and Stephanie E. Smallwood, *Saltwater Slavery: A Middle Passage from Africa to American Diaspora* (Cambridge: Harvard University Press, 2008).

[4] On indenture, see Gaitura Bahadur, *Coolie Woman: The Odyssey of Indenture* (Chicago: University of Chicago Press, 2013). On Indian migration, see Renisa Mawani, *Across Oceans of Law: The Komagata Maru and Jurisdiction in the Time of Empire* (Durham: Duke University Press, forthcoming). For a classic and field-defining account of Indigenous peoples and the Pacific, see Hau'ofa, Epeli, *We Are the Ocean: Selected Works* (Honolulu: University of Hawai'i Press, 2008).

# CONSTRUCTED FUTURES

*Caitlin Mills and Kalina Christoff*

ONE DAY IN OCTOBER 1981, thirty-year-old Kent Cochrane left his job at a manufacturing plant in Toronto and rode home on his motorcycle, as he had many times before. This time was different, however. He lost control of his motorcycle and slid off an exit ramp close to his house. The accident caused a traumatic brain injury, and Cochrane – or K.C., as he came to be known around the world – would spend the better part of the next year hospitalized under medical care. Initially, the extent of K.C.'s brain injuries were not clear. He arrived at the hospital unconscious, and his medical notes indicate that he was having severe seizures. Emergency surgery was performed to reduce swelling on the left hemisphere of his brain. When he regained consciousness seventy-two hours after the accident, it soon became clear that he had suffered significant memory loss: for the first week after surgery, he did not recognize his own mother.[1] Yet it was only after he was released from a rehabilitation facility in July 1982 that neuroscientists began the long process of documenting the true extent of his injuries and their effects on his cognitive brain function. Over the next thirty years, the intensive study of K.C. would lead to profound insights into the functioning of the human brain, the formation of memories, and their links to our ability to imagine the future. When he died, in 2014, K.C. was remembered as one of the most famous amnesiacs in the world.

Initially, studies of K.C.'s injuries focused on his memory loss. In some respects, the patient seemed perfectly normal. He had knowledge about the world, was aware of this knowledge, and could express it easily. If you asked him about the water cycle, he could have explained how water evaporates and then falls back down from the clouds, eventually flowing through rivers into the sea. But he would have no idea whether it had rained the previous day, nor could he recall any specific events that he had experienced personally. In scientific terms, K.C. lacked the ability to form episodic memories, a condition made famous by the award-winning film *Memento*, in 2000. The film tells the story of a man who suffers memory loss after a traumatic event. He tries to re-create the events leading up to his assault and the murder of his wife by using an elaborate system of Polaroid photographs and tattoos to remind him of past events he cannot recall. It may well be that the case of K.C. partially inspired the fictional events of the film. Like the protagonist in the movie, he too apparently relied on notes (left on the refrigerator by his parents and caregivers) to reconstruct the day-to-day details of his life.[2]

As studies of K.C.'s condition progressed, it also became apparent that he had suffered damage to the parts of his brain involved in imagining the future. When asked to imagine some future possibilities – what he might do the next day, for example, or where he might be in one or ten years' time – K.C. could not provide specific answers. For most people, such questions induce a mental *cinéma vérité* featuring themselves as the main character in scenes that include objects and people. For K.C., however, there was only an expansive, prolonged emptiness – no movies of his future self, no imagined scenes that he inhabited, no anchors guiding him into his own future. As he described it to Endel Tulving, a neuroscientist at the University of Toronto, "It's like swimming in the middle of a lake. There's nothing there to hold you up or do anything with." When asked to describe his state of mind while conceiving the future, he replied, between long pauses, "Blank, I guess." He added, "It's like being in a room with nothing there and hav[ing] a guy tell you to go find a chair, and there's nothing there."[3]

Previous to study of K.C., the link between remembrance of the past and construction of an imagined future had not been established in neuroscience. Throughout the 1960s and '70s, multiple cases were documented where brain damage caused patients to lose the ability to remember their own personal past. These patients often suffered damage to the hippocampus, part of the medial temporal lobe in the brain. K.C. – like his famous American counterpart, H.M. – had damage to his medial temporal lobes. Both patients became classic examples of how people can lose their personal memory while keeping other parts of their memory intact. But until K.C.'s case appeared in the 1980s, no one had considered asking patients with medial temporal lobe damage about their conceptions of the future. K.C.'s responses to these questions astounded researchers and set in motion a field of research that continues to this day.

At first glance, the future may present itself in opposition to the past. What happened in the past seems written in stone and unchangeable whereas the future appears to be an unwritten story with limitless possibilities. However, such a framework is overly simplistic when applied to the human mind. Over the last several decades, neuroimaging studies have suggested that remembering our past and imagining our future invoke similar brain functions. Work by psychologists Daniel Schacter, Donna Addis, and others has demonstrated that both memory of the past (e.g., recalling a childhood event) and thinking about the future (e.g., imagining what you will do later this week) activate many overlapping areas of the brain, including the medial temporal lobes damaged in both K.C. and H.M.[4]

Back in the 1980s, the connection between neurological constructions of the past and future was not yet in the forefront of scientists' minds. So, when Tulving asked K.C. questions about how he thought about the future, his answers came as a surprise. "Same kind of blankness," replied K.C., when asked to compare his thinking about what he did yesterday to what he would do tomorrow. Similarly, in the case of H.M., it took decades for researchers to appreciate that the patient's impairment was not limited to thinking about his past but also included an inability to imagine scenarios in his own future – an impairment eloquently captured in the title of Suzanne Corkin's book on H.M., *Permanent Present Tense*. There is something compelling about the notion that remembering the past is in opposition to imagining the future, but this idea is simplistic. Accumulating evidence from patients and neuroimaging studies show that these cognitive functions are highly akin to each other.

One of the biggest advances in the scientific understanding of memory helps us understand why remembering the past and imagining the future are so similar from a neuroscience perspective. This breakthrough was achieved by overcoming the false preconception that memories are fixed like the words in a book, pulled off the shelves of our minds when we need to remember something. We now know that memories, especially personal memories, are reconstructed every time we recall them, mentally simulated out of components of the original experience, rewritten in slightly different ways every time. As it turns out, the same constructive mental simulations produce our thoughts about the future.

The constructive feature of memory has been given a bad reputation by well-known cases of false memories and unreliable eyewitness testimonies. Indeed, the constructive nature of memory can make it less reliable since there is the potential to reshape a memory every time it is recalled. At the same time, however, it is this ability to reshape memory that enables us to learn from and make meaning out of our isolated personal experiences. If we simply filed our experiences like books or essays in the library of our minds, they would remain largely disconnected from one another, unable to tell us what our experiences mean and what our past can teach us about our future. By constructing memories afresh every time we recall them, we also make those memories relevant to our current experiences – highlighting the aspects of a memory that are most significant to our current self while downplaying the aspects that may have been most salient to our past self. Consider, for example, a childhood injury. Early on, the pain of the injury might have been the most central aspect of the experience. As time passed, however, you continued to remember the event as it occurred but likely focused more on what you were doing just before the accident or on who you were with when it happened. The memory of the pain receded to the periphery as the memory was repeatedly reconstructed throughout your life.

Despite their constructed nature, our personal memories are nevertheless constrained by our actual experiences in the world. Are our thoughts about the future similarly constrained? When we imagine the future, are we free to explore infinite possibilities, or is our conception of the future, like our memory, limited by our past experiences? The strong involvement of the medial temporal lobe in thinking about the future suggests that our imagination of the future may be more constrained than previously believed. Scientific evidence suggests that our memories and future projections are built from the same building blocks of personal experiences. As a result, the same forces that can reshape our memories over time – the questions we ask ourselves about our future, our personal motivations while imagining it, or our cultural and social expectations – can also shape and limit our thoughts about our personal futures.

It may seem unthinkable and highly disconcerting to conceive of the future as Kent Cochrane did, surrounded by the blankness of an empty room. But the many parallels between how we remember the past and how we imagine the future point to another much more likely possibility: our personal futures may be heavily (and unknown to us) influenced by the cultural, societal, and personal forces that operate in our daily lives. Although we are gaining a deeper understanding of the flexibility and vulnerability of our personal and collective memories, we have yet to fully appreciate that the same factors may also constrain our ability to imagine the future. Armed with this new possibility, we may be better able to shape our personal and societal perspectives of the future in beneficial ways. For example, how do we ensure that we are not putting unintentional limits on the futures we imagine? Can we alter the experience of children so that, by the age of four, they can have an unlimited number of answers to the question, "What do you want to be when you grow up?"; so that, by the age nine, they will not exclude the possibility of a profession simply because it involves a subject they find difficult in elementary school? The process of reckoning with the malleability of our personal and collective memories has taught us that the past is not written in stone, contrary to what our intuitions may tell us. On the flip side, our personal and collective projections of the future may very well be more limited than we imagined. Becoming more aware of these limits will help us move towards futures we may not otherwise have imagined.

NOTES

1 Shayna Rosenbaum, "The Case of K.C.: Contributions of a Memory-Impaired Person to Memory Theory," *Neuropsychologia* 43 (2005): 929–1021.

2 Canadian Press, "Kent Cochrane, Famed Toronto Amnesiac in Medical Texts, Dies," *CBC Online*, April 2, 2014, http://www.cbc.ca/news/health/kent-cochrane-famed-toronto-amnesiac-in-medical-texts-dies-1.2595665.

3 Endel Tulving, "Memory and Consciousness," *Canadian Psychology* 26, 1 (1985): 1.

[4] See, for example, Daniel Schacter, Donna Addis, and Randy Buckner, "Episodic Simulation of Future Events," *Annals of the New York Academy of Sciences* 1124, 1 (2008): 39–60. See also Karl Szpunar, Jason Watson, and Kathleen McDermott, "Neural Substrates of Envisioning the Future," *Proceedings of the National Academy of Sciences of the United States of America* 104, 2 (2007): 642–47.

# ARTISTIC SILHOUETTES

*Cynthia E. Milton*

ONE OF THE foundational stories in the discipline of art history is that of the Corinthian maid. According to the tale, as told around AD 77 by the Roman philosopher Pliny the Elder, a Corinthian potter's daughter traced her lover's shadow as cast on the wall by the glow of candlelight. The young man, a soldier, was about to leave for war, and his future was uncertain. Moved by the sadness of his daughter, the father made a sculpture of the soldier based on her shadow drawing. This sculpture and the line drawing that inspired it represent traces of a departed loved one, an echo of what was once present.

In late twentieth-century Latin America, where countries suffered the scourge of civil wars, military incursions, and dictatorships, a similar practice of tracing silhouettes of loved ones became common place. Alejandro Obregón's 1962 painting *Violencia* (Violence), which features the outline of a murdered pregnant woman whose body suggests the Andes mountain range, is an early example. Over the following decades, the practice of tracing the outlines of loved ones who had been disappeared, in most cases by the state, grew into the Siluetazo, a political movement demanding their return. Although similar in technique to the shadow drawing of the Corinthian maid, the meaning of the siluetazo representations differs in one important respect. The Corinthian woman made a reproduction of her lover *before* the loss in anticipation of longing and profound sorrow upon his possible death. Yet much of the art in Latin America since the 1970s has been produced *after* a loss. The silhouette outlines of disappeared loved ones that took shape in Argentina subsequently transformed into grainy black-and-white identification photographs representing missing and deceased loved ones across Latin America. These art forms and images are used to remember those who are no longer here. For some, it is the silhouette alone that remains, anchoring the memory of an individual (in the case of Argentina, the memory of a generation) in the present.

Artistic and visual engagements with the past appear to be a near-universal response to violence. In her influential study *The Body in Pain*, published in 1985, social theorist Elaine Scarry suggests that art, in its many forms, may express that which verbal language cannot, helping to achieve a more complete understanding of violent pasts. In a conversation between the historian Gonzalo Sánchez and the artist María Elvira Escallón, whose photography exhibition *Desde Adentro* (From within) commemorated a fire in a Bogotá night club, Sánchez reflects on the limits of written text in recounting the Colombian violence: "A text cannot say everything about the pain that covers our daily tragedies. We need to turn to images, and the multiple possibilities of artistic language."[1]

In Latin America, one of the intended aims of art in response to atrocity is to remember those lost, to protest against barbarity, and to restore the humanity of citizens who have been harmed, displaced, and murdered by the legacies of colonialism and by the slow-burning embers blown into the raging fires of the Cold War. Indeed, the Cold War was far from "cold" in Latin America. Upon his acceptance of the Nobel Prize for literature in 1982, Colombian author Gabriel García Márquez pointed both to the irony of a European jury's fascination with the creativity of the literary genre of magical realism and to their inability to see the surreal reality that had left Latin America isolated: "A reality not of paper, but one that lives within us and determines each instant of our countless daily deaths, and that nourishes a source of insatiable creativity, full of sorrow and beauty, of which this roving and nostalgic Colombian is but one cipher more, singled out by fortune. Poets and beggars, musicians and prophets, warriors and scoundrels, all creatures of that unbridled reality, we have had to ask but little of imagination, for our crucial problem has been a lack of conventional means to render our lives believable. This, my friends, is the crux of our solitude."[2]

In the transition from state violence to democracy in the years following the fall of the Berlin Wall, issues of representation and memory have come to the forefront of political and cultural debates about conflict and repression in Latin America. Earlier protest art that challenged authoritarian regimes and violence has given way to memorial art. While this art emerged from traumatic memories, it is also anchored in the present through its demands for recognition. In Argentina, the silhouettes that stood as silent protests and evocations of missing citizens in the early and mid-1980s now adorn public state-sponsored memory sites, such as the former clandestine centre of the Navy Mechanical School (ESMA). These years were also a period in which a generation of artists from different parts of the world turned to tracing as a means of representing absence rather than presence and, in this way, marking memory and loss.

From 1980 to the mid-1990s, Peru was engulfed by an internal conflict that resulted in over sixty-nine thousand dead and disappeared. For more than a decade, a ruthless Maoist-inspired insurgency group, Sendero Luminoso (Shining Path), fought against violent state armed forces. Both attacked the people they claimed to be defending, and the conflict itself inflamed community tensions. A commission was launched in 2001 to establish the "truth" of what had happened during the conflict and to provide an explanation, if only partial, of why it had occurred. Memorial art flourished in the shade of the truth commission. In one instance (later to be replicated on several other occasions), an art contest was held by a consortium of non-governmental organizations, inviting people from the most heavily affected regions to reflect on the conflict years. The contest was called "Rescate por la memoria" (Recovering/excavating/saving memory), and it resulted in hundreds of works of art, mainly produced by young rural adults (mostly of Indigenous Quechua descent), who used art as a means to remember and recount what had taken place. The images of truth presented in the corpus of works for "Rescate por la memoria" emphasize collective memories of artists' communities, and they articulate individual and shared experiences of suffering. In many of these testimonial images, memories of childhood are brought to light. This art challenges the possibility of reconciliation between those harmed and those who committed the harm in the light of ongoing injustices, and it makes clear demands for recognition of the continued plight of the affected regions. The legacies of the web of violence are intricate and fragile.

Take, for instance, the work *Marks in the Soul: A Girl from Huancavelica* by Rosario Laurente Chahuayo. Using both charcoal on white paper (see next page) and the same image in its negative of white on black, the artist represents a small girl (perhaps a self-portrait). Rosario Laurente Chahuayo, from Huancavelica, would have likely been a child at the time of the conflict. In describing her work, she writes: "Faces of pain and loss of innocent people. In the pain, there is no difference between day or night." This drawing tells a story. The girl stares out with an empty gaze, tears streaming silently down her face; she is inward-looking as she remembers the scene of a massacre that took place in her village. Is this the artist thinking back on when she was a child? Or is this supposed to be a representative child at the time of the conflict, one whose experience would have been familiar to others? In the background, the village is abandoned, dogs are feasting on remains, and the body of a woman lays strewn on the ground (the girl's mother?). In the foreground, we see the child. Her braids are unkempt (she is now an orphan), and her bare feet indicate her poverty. She holds a man's hat in her hands (that of her disappeared father?). The stark palette of white and black has a silencing effect, as though not even a bird is singing. Who committed the violence is, meaningfully, not stated.

Through its ability to bring into a single frame a multiplicity of

*Marks in the Soul: A Girl from Huancavelica*, by Rosario Laurente Chahuayo. Reprinted with permission from Servicio Educativos Rurales, "Rescate por la memoria," 2005.

experiences, art not only helps those who have suffered trauma to shape and give meaning to their past, it also helps those not directly involved to come closer to a sympathetic awareness of the harm done.

Art can express emotions otherwise left unspoken, eliciting an embodied response from the spectator. As author Kyo Maclear has written in the context of Japan after the atomic bombings, art can move viewers "emotionally and intellectually toward the unknown."[3] Art links us to nonverbal memories of the past through the traces of what once was, or might still be, and allows us to share glimpses of those memories with others. For some survivors, art emerges out of a necessity and desire to record what happened for future generations: "Even now [thirty years later] I cannot erase the scene from my memory. Before my death I wanted to draw it and leave it for others," said Iwakichi Kobayashi, a seventy-year-old survivor of the atomic bombing of Hiroshima.[4] Thus, art also asks us to bear witness to the artists' acts of witnessing. It is this connection between *art* and *affect* that may bring us to a closer understanding of the unknown – of something which we ourselves did not experience but which we nonetheless (whether or not directly descendent) have a moral obligation to remember.

Beyond the registers of the past as remembrance and as a present confrontation with our difficult pasts, visual arts may also illustrate a hoped-for future, one of inclusiveness, tolerance, and peace (though not necessarily reconciliation). It is this simultaneous imagining of the future and witnessing of the past that art so powerfully enables. We need to remember – to keep history in our present consciousness – so as to ensure a different future from our many difficult pasts, whether in North America (as in the case of my own country, a settler society), Latin America, or elsewhere. It is this optimism in the aftermath of violence that the charcoal drawing of a girl from Huancavelica holds in common with the shadow drawing by the early Corinthian maid.

These are memories of a remembrance of a lost love as well as gestures towards a hoped-for return and a future promise of Never Again.

NOTES

1. Gonzalo Sánchez and María Elvira Escallón, "Memoria, imagen y duelo: Conversaciones entre una artista y un historiador," *Análisis Político* 20, 60 (2007): 60–90.
2. Gabriel García Márquez, "Nobel Lecture: The Solitude of Latin America," December 8, 1982, https://www.nobelprize.org/nobel_prizes/literature/laureates/1982/marquez-lecture.html.
3. Kyo Maclear, *Beclouded Visions: Hiroshima-Nagasaki and the Art of Witness* (Albany: State University of New York Press, 1999), 24.
4. Japanese Broadcasting Corporation (NHK), *Unforgettable Fire: Pictures Drawn by Atomic Bomb Survivors* (New York: Pantheon Books, 1977), 105.

# MATERIAL PAST

*John Grace and Roland Clift*

ENGINEERS USE SCIENCE to design, build, or change things, ideally to improve the quality of people's lives (which includes not degrading the environment). Unfortunately, much engineering effort is devoted to finding more efficient ways to do things that should not be done at all. Preventing and countering this kind of misplaced effort requires, among other attributes, the ability to remember and learn from past mistakes. In Canada, this historical consciousness is enshrined in the ceremony of the Calling of the Engineer, during which newly graduated engineers are given an iron ring as the symbol of their profession. These rings are thought to have been made from the steel of a notorious, poorly designed railway bridge near the city of Quebec, which collapsed in 1907, killing seventy-five construction workers. The rings serve as physical reminders to recipients of the importance of their duties as professional engineers and the possible consequences of errors in their work.

Memory also plays a role in engineering in a deeper and more profound way. The practice of engineering, and chemical engineering in particular, depends fundamentally on understanding the embodied memory of basic materials at a molecular scale and the expression of this material memory in a larger industrial context.

At the most basic level, the inherent memory of any material is expressed in the chemical bonds that hold its molecules together and in the way those molecules interact with one another to influence the material's structural and behavioural properties. The strength and nature of these bonds vary greatly among different materials and between different phases of matter (solids, liquids, and gases), largely determining how materials respond to external physical forces and chemical stimuli. Many engineering problems require knowledge of the behaviour of multiphase materials – that is, materials that combine substances in two or more physical phases: liquid, gas, or solid. In analyzing these multiphase materials, it is important to understand how they deform or change shape (measured, at a macroscopic level, as strain) in response to applied forces (resulting in stresses). To give an everyday example, when you smear butter onto toast, you are applying a shearing stress. By contrast, if you simply press the butter onto the toast, you are subjecting it to a compressive stress, although it still shears where it oozes out at the sides of the knife. A material is said to be a fluid if it flows in response to small stresses, whereas a plastic does not flow until the stress reaches a sufficient level. Butter is plastic, so it retains its shape after you have spread it, while milk is a fluid and maintains a flat surface in a container or glass. A material, such as rubber, is said to be elastic if it returns to its original shape when the stress is removed.

Viscoelasticity is the property displayed by materials that show a combination of elastic and fluid properties when subjected to stresses. Such materials, often known colloquially as memory fluids, combine behavioural elements of viscous liquids with those of elastic solids. They can thus be thought of as containing an inherent molecular memory. The elastic properties are responsible for the molecular memory, enabling a material to spring back towards its original configuration when an applied force is removed. For example, if you begin to pour a viscoelastic liquid from a jug and then return the jug to its upright position, the liquid may flow backwards, re-entering the container.

Not surprisingly, viscoelastic liquids have many applications in engineering and everyday life. Most synthetic variants of these materials are derived from long polymer chain molecules (such as polyethylene and polystyrene) that account for their elasticity and apparent memory. When stresses are applied to such liquids, deformation may occur because of rapid distortion of the molecules. Such deformation can also occur in viscoelastic solid materials, such as memory foams used in mattresses to provide improved comfort and support. The memory property results in the foam returning to its original shape when stresses (such as the weight of bodies) are removed. These materials have also found widespread use in products such as pillows, shoes, padding, and various medical devices. One special category of memory molecules, known as shape memory polymers, consists of soft materials that have the ability to hold a temporary shape in the absence of an imposed stress. These materials remain in their altered state until they are subjected to a suitable stimulus (for example, change in temperature, chemical environment, or electromagnetic field) that prompts a return to the original conformation. In this way, shape memory polymers have the capacity to store memories of shape for extended periods of time.

Even for fluids that do not show elastic properties, the inherent memory of molecules can be observed in the movement of multiphase mixtures. These memory (or history) effects arise when solid particles, liquid drops, or gas bubbles accelerate through a fluid. In predicting the motion and net drag (resistance to relative motion) force, it is necessary to account for the time delay in the development of the flow around the moving particles and the associated acceleration of the liquid or gas. If the final velocity of a particle relative to the surrounding fluid is sufficiently high, a low-pressure region, known as a wake, develops behind the particle. This effect can be demonstrated by a simple but intriguing experiment. Next time you take a bath, bring a ping-pong ball with you. If you hold it deep enough below the surface of the water and then release it, it will accelerate up to the surface and then stop dead. If, however, you release the ball from a shallower start, say about ten centimetres below the water's surface, it will rise and pop out of the water completely. The first time you see this, you may be surprised: the difference in behaviours is the reverse of what you might expect. The explanation lies in the history of fluid (bath water) flow around the particle (ping-pong ball). When the ball rises from greater depth, there is more time for the low-pressure wake to develop; once it reaches the surface, the force propelling it is cut off. In contrast, a shorter rise does not allow sufficient time for the wake to develop; the buoyancy force acting on the ball is sufficient to propel it out of the liquid. In order to accurately calculate the drag force and acceleration experienced by the ping-pong ball, it is necessary to consider and remember the history (that is, the development) of the fluid flow surrounding the accelerating ball.

For many practical cases – such as particles accelerating in gases, which have a density that is orders of magnitude less than that of the particles – history effects are negligible. On the other hand, memory effects can be highly important when the fluid and particle densities are similar, as in the case of particles in highly turbulent liquids, such as vigorously boiling water. Accounting for the development of the flow field is also important in some industrial processes – for example, when aluminum particles are injected into molten steel to react with and remove oxygen. In this case, the history effect has a large influence on the length of time that the aluminum particles remain in contact with the molten steel and hence on their ability to capture and remove oxygen. The resulting oxygen content, in turn, influences the properties of the resulting steel and its suitability for various applications.

Moving beyond molecules, bath tubs, and industrial furnaces, memory effects are also critical when we seek to manage the environmental impacts of material products. In particular, it may be important to account for the past uses of objects when considering their future fate and whether they should be reused, recycled, or "downcycled" into a different product or discarded. To this end, we need to determine the full environmental impacts (or footprint) of the industrial activities making up the supply chain delivering products or services. The tool used for this environmental analysis is known as "life-cycle assessment." The underlying idea is simple: examine, evaluate, and sum the environmental impacts of all the activities leading up to the product and using it and managing it after use. But, as with so many things, the devil is in the details, and the assessment is, in practice, often complicated and time-consuming. How does one, for example, account for the environmental impacts of a material that is used several times in very different products? Should one regard a secondary material, recovered for recycling, as free from the environmental impacts of previous uses or insist that the cumulative history of the material be remembered in accounting for its next use? The distinction matters insofar as it encourages differing approaches to reusing material products. Treating secondary materials as free from inherited environmental burdens (that is, without an embodied memory) tends to encourage recycling, whereas remembering past impacts encourages reducing material usage in the first place by, for example, extending the lives of products already in use. Both approaches are desirable in seeking more sustainable ways of living, but the preferred option in any given case depends on circumstances and personal convictions and priorities.

Steel provides an important and interesting example. This material has supported (in both a literal and figurative sense) the development of the industrialized world. It is a material that can and should be recycled or down-cycled and used numerous times in many different types of products. However, steel has a range of compositions that determine how it is recycled or down-cycled and the applications of the resultant product. In the case of building construction, there is debate over whether steel or reinforced concrete should be preferred on environmental grounds. This is not a straightforward question: answering it requires assessment of the life cycles, not only of the steel or concrete but also of the building itself. Recycled steel has a much lower environmental footprint than virgin steel (steel initially forged from elemental iron, nickel, other metals, and carbon), but the supply of recycled steel may not be sufficient to meet demand. Some in the steel industry have suggested that because steel can be recycled, all steel should be treated as if it is recycled, because it will, ultimately, be recycled. This argument, which has not been widely accepted, would make steel appear to be more sustainable relative to other materials, amounting to a kind of false memory.

Memory also figures in the choice between reusing or recycling steel. Certainly, reusing a steel component, such as a beam or joist, is even better than recycling the metal. However, it is necessary to know the prior use of the steel beam to understand its suitability (in terms of strength, flexibility, and other properties) for a repurposed use. For example, steel beams from old buildings often have to be reused in places where they are unnecessarily heavy and strong. The original specification of the component (its industrial memory) is thus critical to ensuring its appropriate reincarnation.

Accounting for the environmental impacts of industrial processes and products amounts to evaluating and allocating guilt for past ecological sins. Opinions differ on whether these sins can be completely forgiven and forgotten in the subsequent lives of the material in question. Among life-cycle-assessment practitioners, different approaches to this question are really a matter of convention in accounting for environmental impacts, but the difference has provoked hot and unresolved debate. The dilemma illustrates that engineering, far from being a straightforward application of science to design and technology, inevitably crosses over into issues and questions that involve the culture, social conscience, and sometimes even the religious beliefs of the practitioner. There is no technical answer to inform decisions on freedom from past environmental burdens. Engineers, then, like medical doctors and lawyers, must confront questions that are linked to culture and hence to the past, even in issues that may appear to be purely scientific in nature.

# CRITICAL PERIODS AND EARLY EXPERIENCE

*Janet F. Werker and Lawrence M. Ward*

MUCH OF OUR EXPERIENCE in the world is shaped through the lens of our memories. We can establish memories – albeit at different levels and of different kinds – across the life span. Yet not all these memories are created equal. There are windows of time – critical periods – when the human brain is particularly open to specific types of experience that lead to a lasting imprint on neural organization and perceptual and cognitive functioning. For example, critical periods in the development of face recognition or sound-processing pathways allow us to learn the specifics of our ecological niche, including the faces and language of our "tribe," so that we can function efficiently and optimize survival during adulthood. Critical periods reflect memory in the double sense of the word – both in the lasting impact they have on the development of individuals and as a signature of the evolutionary history of our species.

From a biological perspective, a critical period is a time window during development when a particular neural system is unusually malleable in response to a particular type of experience. Over the past several decades, we have begun to understand the neurological basis for learning and memory formation during critical periods. We know, for instance, that parvalbumin neurons act as inhibitory regulators. By mediating communication between other neurons, they control the timing of critical periods. When the ratio of inhibition to excitation in these neurons reaches threshold, a series of neurochemical events occur that enable the connections with the other neurons to be rewired into a functional circuit. This neural rewiring plays a critical role in learning, and it is not unique to critical periods. But what distinguishes critical periods from ordinary learning is the existence of neurochemical processes – so-called molecular brakes – that stabilize the new connections and inhibit further learning once a critical period has ended. Outside of a critical period, when the brakes have been applied, learning relative to the critical experience is much more difficult and, in some cases, impossible.

The molecular mechanisms that underpin critical periods are similar for many types of sensory learning and have remained nearly unchanged across the evolution of vertebrates, from reptiles to humans. David Hubel and Torsten Wiesel's work with cats in the 1970s illustrates this. Their now-famous experiments showed that if one of a cat's eyes is deprived of visual input early in life (they sutured the lid closed), the other eye takes over, projecting its sensory information to areas of the visual cortex that would typically be activated by the incapacitated eye. The longer the deprivation, the greater the effect, and the more difficult it would be to reverse (by removing the sutures from the cat's eye, for example). This limitation can have serious consequences for vision, making it difficult to use binocular depth cues to guide spatial navigation, sometimes resulting in accidents.

In humans, each critical period, including the well-defined one that shapes speech perception, has cascading and long-term effects. Consider, for example, the acquisition of language in babies. During the first year of life, babies become experts at discriminating speech sound distinctions used in their native language, but they become worse at discriminating those not used. In 1984, Janet Werker and Richard Tees reported in "Cross-Language Speech Perception" that an English-learning infant will show, by the age of just ten to twelve months, a decline in its ability to discriminate two different *d* sounds used in Hindi, while a Hindi-learning infant maintains and sharpens this ability. In 2005, in "Infant Speech Perception Boostraps Word Learning," Werker and Henry Yeung reported further on the impact of this phenomenon on word learning. They discussed, for example, how an English-learning toddler will expect the words *doll* and *ball* to refer to two different objects, whereas a Hindi-learning toddler will expect *dal* (lentils), produced with a dental closure, and *Dal* (branch), produced with a retroflex closure, to refer to two different objects – as they do in Hindi. Given this knowledge, there has been an increased focus on hearing screening in infants in order to identify and correct potential problems and avoid long-term difficulties with language acquisition.

Critical periods can have significant negative effects on language learning, particularly in the case of second-language acquisition. For immigrants, high fluency in the language of their adopted country predicts income, job success, satisfaction, and other social determinants of well-being. In a 2015 article published in *IZA World of Labor*, Ingo E. Isphording showed that immigrants to a new country who are older than about eleven years typically have less success in learning a second language, particularly a language linguistically distant from their native tongue. Because culture is transmitted in humans mainly by language, difficulties acquiring the language of an adopted country could have significant implications for the cultural integration of immigrants.

The functional benefits of critical periods are evident in other areas. We can appreciate the social benefit of learning the musical rhythms of one's native culture – a process that has a critical (or at least sensitive) period in early infancy, as Erin Hannon and Sandra Trehub demonstrate in their 2005 study, "Tuning in to Musical Rhythms." But critical periods can sometimes have negative effects as well. Being exposed in infancy only to the faces of people with the same ethnicity can have long-term negative consequences for interactions in a diverse culture. Infants exposed only to Caucasian faces early in life, for example, are less able to tell two different African faces apart. In a 2007 study, "The Other-Race Effect Develops during Infancy," David Kelly and colleagues showed that someone exposed only to Chinese faces will have more difficulty discriminating among Caucasian faces. The consequences for misidentifying a criminal in a lineup are obvious, and the implications for justice and racial inequality are serious.

Not all critical periods occur early in development. Daniel Levitin reported in his 2006 book, *This Is Your Brain on Music*, that our musical preferences are typically shaped by the genres we listened to in our late teens or early twenties, hence the proliferation of classic rock radio stations in North America that cater to the baby boom generation. It is rare for musical preferences to change in adulthood, and fairly significant social upheaval or immersion (think the 1960s) is required to bring about a deep and lasting change. Is this kind of cultural preference the same as the early-appearing sensory and linguistic ones? Does it involve the same parvalbumin interneurons or a different set of interneurons active in more slowly developing parts of the brain? Perhaps it involves a different mechanism entirely? These questions have yet to be answered.

Other lasting effects of cultural socialization, including effects on so-called higher-order psychological processes such as decision making, may also be developed via critical periods of varying intensities. Take empathy, the ability to "put yourself in another's shoes," for example. In 2008, in "Putting the Altruism Back into Altruism: The Evolution of Empathy," Frans de Waal theorized that empathy is the basis for directed altruism, leading to selfless actions in response to another's pain, needs, or distress. Does empathy for others have a critical period? A 2013 study by J. Xavier, E. Tilmont, and O. Bonnot, published in the *Journal of Physiology, Paris*, shows that there is, indeed, a developmental trajectory: the capacity for empathy develops until about the age of six or seven. Does the capacity for empathetic learning then end before puberty? And is empathy restricted only to those people who have the faces of, and speak the language of, a person's native tribe? These are important questions.

Religion is another type of cultural learning that has lasting effects on perception, thinking, and beliefs. Most people in the world today are raised within a specific religious (or explicitly nonreligious) context. Analyzing international surveys from the 1990s and early 2000s, John Barro, Jason Hwang, and Rachel McCleary reported in a 2010 study, "Religious Conversion in 40 Countries," that relatively few people stray from these early influences: between 80 percent and 98 percent stay in the religion (or lack thereof) of their childhood until they die. Even those who later eschew religion altogether continue to be strongly influenced by the religious traditions of their early lives. The researchers found that religious conversion happens most readily when people are exposed to a variety of other religions in a diverse culture, as in Canada or the United States. But even in those countries, religious conversion is atypical. Babies are not born knowing the traditions of the religion to which their parents subscribe; they must be taught their "birth religion." Perhaps religious teachings exert such a strong influence because they typically occur early in development. Are we to conclude, then, that there may be a critical period for the development of religious belief? The fact that only 20 to 30 percent of people will convert from atheism to religious practice in spite of exposure to a dominant religious culture suggests that there might, indeed, be a critical period, albeit an imprecise one.

Some critical periods have abrupt onsets and relatively sharp offsets, whereas others are more gradual and extend over longer periods of time. It may also be that the timing and duration of critical periods is more flexible than previously believed. We once believed that a critical period, when closed, could not be reopened, and we interpreted the existence of learning after the closure as evidence that the critical period had not existed at all. In the area of language learning, however, a lucky few people do succeed in learning a new language fluently and without an accent after puberty. This does not mean that language learning in these individuals is not subject to a critical period. Rather, we now know that the timing of critical periods can be accelerated or delayed and that critical periods can be kept open beyond the normal time period.

Critical periods can also be reopened later in life. In 2013, Anne Takesian and Takao Hensch, in "Balancing Plasticity/Stability across Brain Development," reviewed a number of studies showing that the external delivery of pharmacological agents or events that lead to the internal release of neurochemicals (both of which act on the parvalbumin cells) can prematurely open, keep open, or reopen neural plasticity. This striking finding raises a wide range of biological, cultural, and ethical questions around the possibility of "engineering" a reopening of critical periods to facilitate learning and memory creation outside of the biologically programmed window. We could, potentially, correct a number of problems: lifelong deficits in language and reading among children born with hearing impairments, the facial discrimination difficulties of children born with cataracts, and language-learning difficulties among adult immigrants. The benefits are obvious, but what are the costs?

Several recent examples offer a glimpse of what the manipulation of critical periods might look like. In 2013, Judit Gervain and colleagues reported in *Frontiers in Systems Neuroscience* that the administration of the drug valproate, along with intense auditory training, can improve absolute pitch perception in adult human males who have long passed the critical period for such learning. A study published in *Neuron* in December 2012 likewise reported that pharmacological interventions that influence NMDA receptor regulation (and, indirectly, parvalbumin cells) can delay the loss of, and help regain, patterned vision in mice that have been genetically modified with impaired eye muscles (a symptom of Rett syndrome in humans). Clinical trials are now underway in humans. In the future, similar approaches might conceivably be used to reopen critical period plasticity for adults who are moving to a new country and need to acquire an additional language or for people such as child soldiers who did not have the opportunity to learn empathy from a young age. Although these potential interventions raise hope for a better future, it is easy to imagine the horrors that could arise from their misuse. We must proceed with caution if we intend to tamper with deeply embedded memory mechanisms.

Critical periods help humans adapt effectively to their ecological niche. Biologically evolved brain mechanisms control the opening and closing of these learning windows, and we are discovering how to manipulate these mechanisms to prolong the benefits and ameliorate the costs of their operation. As the lives of humans become more complex and culturally evolved, it is interesting to consider how the functioning of critical periods might adapt to meet the demands of our new multicultural, multilinguistic, and technology-dominated environment.

# RELEASING TRAUMA

*George Belliveau*

PLAYWRIGHTS AND PERFORMERS make use of memory both to blur and to enlighten past and present events. Characters can play out moments from the past to re-examine their history and gain a deeper understanding of the present. As seen and performed through a theatrical lens, memory is literally embodied in the characters onstage, providing a unique vehicle for expression and understanding. In the face of crisis, playwrights, and sometimes actors, can use memory plays as a form of psychological recovery from trauma. In a theatric retelling, artists can look back at what happened, contextualizing the traumatic events within their life story while using creative expression as a form of coping and healing.

Tennessee Williams' *The Glass Menagerie* is often described as the first English-language play to bring the genre of memory play to the forefront. The main protagonist, Tom Wingfield, begins the play by directly addressing the audience as he proceeds to guide them through his family memories. In this way, Tom relives moments from his past while simultaneously existing in the present moment on stage. Many playwrights since Williams have explored this theatrical device. Canadian playwright Sharon Pollock makes use of multiple first-person memories in *Doc*. In the play, recollections of the past emerge from more than one character or perspective, challenging the notion of an absolute truth. French Canadian Michel Tremblay pushes this technique even further in *Albertine en cinq temps*, in which the lead character, Albertine, engages her past self at five different moments in her life, as portrayed by five different actors.

Playwrights are increasingly writing one-person memory plays in which actors can use their past to inform the present (or *now*) of the play, as the present provides a richer context to understand past events. Joan McLeod's *The Hope Slide* is a one-woman play in which the action moves across time, from the play's present to events remembered from the past. In it, as in Pollock's *Doc*, McLeod introduces memories to reveal tragic events that shaped the protagonist's present-day realities. Tetsuro Shigematsu's *Empire of the Son* layers the use of memories in similarly reflexive ways. As a playwright-performer, he plays both himself and his Japanese father. The two characters exist simultaneously in the play's present and in a complex remembered past.

Memory plays often depict difficult moments and past memories of trauma with which playwrights are still grappling. For example, the memories depicted in McLeod's *The Hope Slide* include a series of untimely and unnecessary deaths in the small town of Hope, British Columbia. A man dies in a hunger strike, and a woman falls victim to a landslide. The trauma faced by this rural community is symbolic of McLeod's community of artists, who at that moment in history, the 1980s, were dying from AIDS. In *Doc*, Pollock revisits her adolescent past, where she lived with a dysfunctional family, including an alcoholic mother, Eloise, and a workaholic father, Ev. Eloise's lack of communication with her husband, combined with long-term depression, led her to commit suicide, leaving her teenaged daughter to cope with many unanswered questions.

My own connection to the memory play goes beyond an academic interest. At a deeply personal level, I too have used this form of theatre to explore a traumatic memory – the unexpected and tragic death of my brother, Don. What follows is an excerpt from a memory play I wrote and performed called *Brothers*. This solo piece explores how my memories of growing up with Don in rural French Canada have shaped our identity and destiny. It provides an example of how theatre can be used to explore complex and difficult memories, enabling past and present to simultaneously come to life and inform each other.

---

## *Brothers*

[*Set consists primarily of two wooden blocks about three metres apart. George sits and stands on them.*]

GEORGE

(*Enters and manipulates scarves on the floor.*)

Brother. Don

Strong, smart, stubborn
While all other ten-year-olds played hockey and team sports
My brother, Don, took his snowshoes to the forest to check his traps
Spent hours in the gym, disciplined, poised
Sought new challenges, new horizons
Ice climbing, rock climbing, mountain climbing
He climbed in groups, as a guide, and often alone

His younger brother, George (that's me)
Outgoing, chatty, social
I loved hockey and all team sports
If I wasn't playing hockey on the ice, I'd play in our garage
Sometimes with friends, often on my own
Playing for hours
Play by play, broadcasting each move
(*Picks up a scarf and uses it as a hockey stick.*)
"Belliveau, prend la rondelle. La passe a LeBlanc. C'est le but!"
I made my way to all the nooks and crannies of our garage
In my element, playing

Don and I grew up in French
A small Acadian village in eastern Canada called Memramcook
An Aboriginal name for "crooked river"
When Don was ten and I was nine, our friend Luc slipped in that winding river
It was the first funeral I ever attended
Each time Don and I crossed the bridge over the meandering river

(*Holds his breath as he steps over the river, pauses, then releases the breath.*)
We never talked about why we did that; we just did
In our teens, we started to listen to English radio and television
Stumbled through English newspapers and magazines
Met friends from the neighboring English town
An den' wit our tick French accents, we set off to Université
En Anglais, in English
Don fell in love with robotics and me for the theatre
When not studying, he climbed, and I played hockey
From hockey to theatre dressing room I travelled
From lab to mountains he journeyed
I was wrestling with Shakespeare, butchering the language
(*Climbs onto block.*)
He was staring at the Himalayas, clenching for dear life
Don summitted Mt. Tilicho, one of the great peaks in the world
As Duke Orsino in *Twelfth Night*, I left behind my French accent
(*Jumps off the block.*)

## MOM

(*Played by* GEORGE. *She is folding laundry. She listens, and then she chuckles to herself.*)
He doesn't know I can hear him mumbling the play-by-play through the wall! That George, my youngest, he always liked to talk. It's like he has his own stage and hockey rink in the garage! When he gets going, you'd think he has a whole team out there ... he plays for hours!

His brother plays with him once in a while. But Don'd rather be out in the forest. You really couldn't have two different boys. One can't stop talking, and the other hardly says a word. Still, they're good friends. Take care of each other.

When they come home from Université, they share the same room, even with two other bedrooms sitting empty. They talk and laugh until the wee hours in the night. Bright and early, they head off together, Don to the gym and George for a run.

DON

(*Played by* GEORGE.)

It's hard to describe exactly what happened. I was climbing, taking in the view from the top of the world, up about four thousand metres, touching the clouds really, then all of a sudden I was at three thousand.

(*Beat.*)

I remember sliding, trying to hold on to my pack.

(*Looks up.*)

The moon is beautiful.

GEORGE

I was playing the director in Pirandello's *Six Characters* – a demanding role. I was doing another play at the time and finishing my teaching degree. Fully sleep deprived. It happened in Act II. The six characters are trying to convince me that their story is worthy of the stage. They are there, and they are not. I'm there, and I'm not. I'm drifting into another world.

DON

How long have I been sliding? Thirty seconds? Thirty minutes? Just breathe in this stunning vista. Valleys, peaks.

GEORGE

I don't know my lines. They're gone.
(*Lost, so improvises the following lines to imaginary actors.*)
"Okay, I'd like you to move towards Madame Pace's dressing area. And you, you need to stop crying." I have no idea where I am. How long have I been lost? Five seconds? Five minutes?

DON

(*Waking up awkwardly. In awe.*)
Just look at the sun creep over the mountains, filling in the valleys. I've never seen a dawn like this. Am I awake, dreaming?
(DON *touches his cheek, winces, struggling.*)
Ooh, my shoulder. Where's my backpack?

GEORGE

(*Softly, as an actor.*)
"Mr. Director, can we show you the scene between the father and daughter?"
(*Back as the director character.*)
Thank you.
(*Finding his place in the script again, addressing an imaginary actor.*)

"Sir, this confession tale between you and your daughter is touching, but it won't play on stage. What we do here in the theatre is make-believe."

## DON

(*In distress.*)

Once in your life, George, you have to see the sunrise in the mountains. It's absolutely beautiful.

## GEORGE

Don's passion and determination to test his limits, push the boundaries, live each day fully never ceased. Carpe diem!

The second time he slid, in the rugged Canadian Rockies, among the beauty of snow, rocks, and endless sky, he wasn't so lucky.

Now, when I get lost, when I start to slide, he gives me a push ... or a kick in the butt, helps me see the beauty and find my way.

We're still connected.

---

The memories in this piece were initially generated by photos, videos, artifacts, and conversations with friends and family as I tried to recapture the past. The poetry of the piece came to life, however, through editing, rewriting, and letting go of moments that were too descriptive. I sifted through language to find the essence and spirit of these memories. But the vitality of the piece only really emerged in the theatre space, where the memories became three-dimensional and were given a pulse. The words on the page shifted to a living, breathing remembered past. While rehearsing and performing, forgotten aspects of the past would sometimes take me by surprise; in the act of performing, I was able to release memories that I had previously failed to recall or, potentially, process. This notion of releasing memories through creative or embodied work aligns with the way traumatic experiences are processed. For instance, in rehearsing the narrative about crossing the bridge where our friend Luc had drowned in the river, I broke down and could not speak for several minutes. The only way to give language and truth to that moment was to take a breath. I cut the narrative text and replaced it with a stage direction for the actor to "hold his breath, pause, then release the breath." Crossing that same bridge recently (forty years later) while revisiting my hometown, my instinctual reflexes had me once again hold, then release, my breath.

Although memory plays such as *Doc*, *The Hope Slide*, *Empire of the Son*, and *Brothers* are about traumatic moments from the past, they also serve a recovery function or as a way of coping for survivors in the present moment. The duality of the text, consisting of events from the character's past at the centre surrounded by a secondary framing text played out in the present, is a reflective testimony of the complex, often unresolved past. A further layering exists in plays performed by the playwright, as the lines between character and actor are blurred, adding to the emotional complexity. Audiences who witness memory plays (whether performed by the playwrights themselves or not) are brought into a dynamic interplay between past and present, seeing both the historical trauma and the way that individuals cope in the present and try to move forward. Within this interplay of time and space, audiences make meaning and often experience a measure of catharsis. Witnessing others live and overcome trauma allows a release, a letting go of what audiences might be holding on to as they enter the theatre.

# A FISHY STORY

*Anthony Farrell*

HAVE YOU EVER FORGOTTEN your way home? As a young lad in the United Kingdom, I tried very hard to do so. Most weekends, I took a bus to the nearby countryside to wander through farmers' fields and forests, leaving behind the pollution, noise, and dirt of my industrial town in the Black Country. Many hours later, I returned to the bus stop to begin my trip home. Uncannily, I always managed to find my way back despite never carrying a paper map or a compass, let alone a cell phone GPS, which was still decades away. Clearly, I wasn't orienting in a traditional way.

Now, after travelling to every continent, I am acutely aware that I can visualize maps in my brain, even if I have only visited a place once before. Perhaps this helps explain my interest in salmon migrations, developed over a decades-long career as a fish physiologist. Like the younger version of myself, learning the way home is perhaps the most important aspect of memory for these magnificent migratory animals, whose travels across wide expanses of the oceans must surely constitute one of the wonders of the natural world.

The ability to navigate home is a crucial memory that must be laid down early in a salmon's life. Consider the case of a Pacific pink salmon newly hatched from its egg somewhere along the coast of the Pacific Northwest. Conceived only a few months earlier and feeding on the remnants of its egg yolk, it weighs less than half a gram when it emerges from the semi-dark of the gravel stream bed in search of food. Undaunted by its diminutive size, it embarks on a swim of some ten to a hundred kilometres downstream to the sea, where it will exploit the abundance of oceanic food during an almost two-year walkabout before a return migration. After reaching its natal spawning area and mating, it dies naturally within a matter of days. Our salmon can't afford to get lost on its way home – there will be no second chance to spawn. For this salmon, learning and remembering the way home is essential to propagating the species; its memory has been shaped by strong evolutionary forces. For salmon species that spend longer (up to five years) growing at sea, the challenge of remembering the way home is even greater.

How does an adult salmon find its way home from the open ocean, undertaking a journey of up to several thousand kilometres, for the first and only time, with remarkable fidelity? We cannot talk to salmon, nor can we perform controlled experiments with wild fish travelling over huge distances. So how have scientists reached the conclusion that long-term memory is needed to get home? One part of the reasoning involves making sense of the features of salmon migrations that are shared across individuals and species. Although a given population of juvenile salmon may migrate downstream together, kinship is relaxed and groups disperse over large areas once they arrive at the open ocean. Indeed, salmon species and populations (from both the Asian and North American continents) intermingle as they search for food in the offshore waters of the Pacific Ocean with the result that individual fish from a wide variety of home locations experience similar environmental conditions (current, temperature, salinity, food resources, and so on) during their time at sea. Yet despite this mixing, representatives of each salmon population will independently converge on their respective home rivers with a precise timing that is specific for each population. There is clear evidence that the swimming directions of salmon in the ocean (and in a river) are not random. Adult salmon can thus orient at sea and reach their natal river at a scheduled time, despite being spread over vast areas. The ability of salmon to orient in both space and time is also inferred from the fact that all fish travel at about the same speed (maximizing energetic efficiency), irrespective of their starting and ending points. This requires fish farther away from their spawning grounds to leave sooner in order to end up in the same place at the same time as their kin. Collectively, these observations strongly suggest that the salmon's homeward migration is guided by some form of spatial memory.

If memory is, indeed, critical to the homing migration of salmon, what information is used to form these memories? How is it, for example, that a juvenile salmon can memorize the specific piece of coastline associated with its natal river? Many migrating animals – including pigeons, salamanders, and sea turtles – use compass orientation, but a compass is insufficient for navigation on the open ocean, as it only provides information on heading or direction. For this information to be useful, it must be accompanied by a map to provide a geospatial reference. Scientists believe that salmon memorize some form of map when they migrate out to sea as juveniles and use this map as the basis for their precise return migration. Research over the past several decades has shown that salmon use a bicoordinate system of navigation at sea that enables them to know their location, where they need to go, and when to leave to get there on time. This navigational road map is based on natural variability in Earth's magnetic field. By memorizing fine-scale features of the Earth's magnetic field, these fish are able to locate the precise magnetic coordinates of the location where they entered the ocean as a juvenile. Adults orient to these coordinates when they return to spawn.

Salmon navigational systems are not perfect, however. The strength and inclination of Earth's magnetic field varies from year to year, and it may not be exactly the same on the return journey as on the outward journey, when the memory was imprinted into the juvenile fish. This variability is thought to explain interannual differences in the migratory routes of Fraser River adult sockeye and pink salmon around Vancouver Island. Although these fish tend to swim in a general east-southeast direction from their feeding grounds in the Subarctic Pacific Ocean (as opposed to Asian populations that swim west), there are multiple routes they can take when they return to the Fraser River, entering the Georgia Strait from either the north (through Queen Charlotte Sound) or the south (through the Strait of Juan de Fuca). Salmon surveys over many decades show that a higher proportion of sockeye salmon migrate through the northern route when the difference in magnetic intensity between the Fraser River and the Queen Charlotte Strait decreases. Conversely, a higher proportion of salmon migrate through the southern route when the difference in magnetic intensity between the Fraser River and the Strait of Juan de Fuca decreases. While these findings are consistent with predictions for magnetic imprinting, other factors, most notably temperature, also affect the oceanic swimming direction of adult Pacific salmon. Moreover, since daylight, rainfall (which affects sea surface salinity), and tidal cycles all vary in a predictable fashion with latitude, adult salmon could potentially use these environmental signatures (once again imprinted as juveniles) to guide their return home. These ideas have yet to be experimentally tested.

The navigational challenges of salmon migrations do not end when the fish reach freshwater. They must still find the particular location (a stream, creek, or lake) where they first hatched into existence. Many experiments have shown that this freshwater phase of the homeward migration relies heavily on olfactory information (memories) learned during their seaward migration years earlier. As a juvenile, a salmon memorizes odours at one or more locations and stages of their lives in freshwater, taking advantage of variability in the chemical characteristics of a river caused by the natural heterogeneity of rocks, soil, and plants. In all likelihood, salmon commit to memory a series of olfactory waypoints at the confluence of each new river they encounter on their way to sea. As adults, they are attracted to these odours on their return home.

Fraser River sockeye salmon decrease their rate of travel from about twenty-five kilometres per day to about ten kilometres per day once they reach the southern Strait of Georgia, close to the mouth of the river. The time spent in the vicinity of the river helps the fish adjust to new environmental conditions (salinity and temperature) and also, in some populations, to reach an appropriate stage of sexual maturation. During this waiting period, salmon segregate in space and time, according to their river of origin. Kin recognition in these mixed groups of populations and species involves olfactory memory imprinted during juvenile life stages. By increasing their vertical movements through the water column and periodically surfacing into fresh, odor-rich waters, the salmon are able to "sniff" the olfactory environment of the near shore waters.

The concept of juvenile salmon being attracted to learned odours is well established (concepts of imprinting were developed to a large extent by the famous ethologist Konrad Lorenz). For example, juvenile salmon released upstream of (rather than at) the location where the eggs hatched will return as adults to the new location. Fish released from hatcheries can be imprinted with artificial chemical odours to return to a non-natal hatchery. Although there is no doubt that adult salmon have memorized olfactory waypoints for freshwater navigation, other factors may "tune" the migration. For example, an excessively high river temperature may cause fish to hold stationary in lower river reaches or lakes, awaiting cooler water. A similar holding behaviour occurs if the river velocity is too high for passage.

Once a return migration has started, time is of the essence. Indeed, most adult salmon in rivers have just weeks before they mature and die naturally. This innate clock – a biological ticking time bomb – cannot be stopped. Consequently, salmon have an imperative desire to reach the natal spawning area – hence their remarkable swimming feats while negotiating waterfalls and rapids. The timing of each migration event is predominantly an evolved (genetic) adaptation to the prevailing environmental conditions encountered along the migration corridor and spawning grounds. The event is genetically controlled because evolution has favoured salmon migrating at the right time of year for river temperature and flow, as well as for breeding and the subsequent hatching of offspring during periods of food abundance in spring. The homeward migration must be initiated months ahead of time, with no advanced information on future river conditions. Salmon need to reach the spawning grounds with enough energy to spawn, often after having swum thousands of kilometres through the ocean and against raging rivers. This feat is all the more remarkable when we consider that sockeye salmon will stop feeding up to months ahead of spawning and may swim as far as one thousand kilometres upstream over several weeks without stopping to take on fuel.

The idea that animals can migrate with geographic precision and timeliness has long fascinated humans, and the need to understand the patterns of these migrations is critical for the success of those who depend on the harvest of animal products. Indeed, predicting the location and timing of salmon returns into coastal and estuarine habitats has become essential during centuries of ceremonial, recreational and, most recently, commercial fishing. The ability of Pacific salmon to learn and remember their way home is a remarkable natural event, even more remarkable, perhaps, because it does not require the cortical complexity of the human brain.

All the same, one should not be left with the impression that the salmon migration is perfect. If this were the case, salmon would not abound in British Columbia today. The present-day salmon of British Columbia are postglaciation invaders, relatives of those that strayed from more southern, ice-free natal spawning areas following the last ice age, between fifteen thousand and ten thousand years ago. Even today, a small percentage of migrating salmon (between about 1 and 10 percent) will deviate from their intended route to return to a different spawning ground. Straying may increase when migratory river conditions are unfavourable or when fewer salmon return to a river. These "lost" fish play an important role in the genetic exchange of wild salmon populations and in the distribution of fish across all potentially suitable habitats.

Like some members of the salmon populations of the vast Pacific Ocean, humans too can stray. I, for example, ventured across an ocean and a continent to settle on the west coast of North America. My father was one of the lucky ones to return home to the United Kingdom from the Second World War, having served in the Pacific, Indian, and Atlantic Oceans and being sunk while his ship patrolled Algiers Harbour. He remained restless in the United Kingdom after the war, wanting to stray to another country, but in the end remained faithful to his home. He left the migration to his offspring, who did, eventually, forget his way home.

ACKNOWLEDGEMENT

I would like to extend my sincere thanks to Tom Quinn for generously allowing me to read ahead of print some relevant chapters from his revision of *The Behavior and Ecology of Pacific Salmon and Trout*.

# RECONSTRUCTING THE PAST

*Megan E. Giroux, Lara Boyd,*
*Deborah A. Connolly, and Daniel M. Bernstein*

On August 1, 1984, Ronald Cotton's life changed forever. This was the day he was arrested and charged with burglary and rape. The primary evidence against him was an eyewitness identification made by Jennifer Thompson, a twenty-two-year-old college student who was the victim of these offences. Thompson claimed that she had studied her assailant's face, trying to remember every physical feature so she could later identify him. She subsequently selected Cotton from a series of photographs and, later, from a live lineup. In January 1985, Thompson confidently testified against Cotton at trial. Cotton was convicted on both counts and sentenced to life in prison. He would spend the next ten years of his life in prison for a crime that he did not commit. In 1995, DNA evidence demonstrated that the crimes he was alleged to have committed had actually been committed by Bobby Poole. A decade earlier, Poole, who was an inmate at the same prison as Cotton, had confessed to these crimes. Yet, at a retrial in 1987, Thompson claimed to have never seen Poole in her life. Erroneous eyewitness identification thus led to the conviction of an innocent man.

Legal systems around the world rely on human memory, whether it be the institutional memory of precedent, from which common law is constructed, or the personal recollection of past events in witness testimony. In civil and criminal cases, witness memory is often used as crucial evidence and may be the primary or sole evidence of an accused's guilt. Such reliance forces us to critically evaluate the limitations of memory and the circumstances under which it may be unreliable. Memory can be, and often is, highly accurate, and it can form the basis for sound legal convictions. Yet any particular memory is also malleable – existing on a continuum from accurate autobiographical memory to distorted and false memory. Distorted memories reflect corrupted recall of details of events that we have experienced, whereas false memories represent recollections of entire events that we never experienced – fictitious stories cast as autobiographical documentaries in our minds.

The occurrence of distorted or false memories can be understood from the manner in which memories are formed and stored in the human brain. At the most basic level, the brain must perform three main functions to create a memory. Information from a lived experience must first be encoded. This information must then be consolidated, or stored, in the brain. Finally, when needed at a later time, information must be retrieved for use in new and different contexts. From a neurobiological standpoint, encoding of memories occurs through increases or decreases in chemical signalling between neurons, which can happen rapidly during the formation of short-term memories. However, the information captured in these short-term memories may be lost if it is not converted into long-term memories. This conversion to more permanently encoded memories, the "consolidation" phase, requires structural brain changes, including alterations to the number and complexity of the terminal branches on neurons (called dendrites) and strengthening in the physical connections within and between brain regions. These structural changes take time (somewhere between four and twenty-four hours), likely require sleep, and benefit from repetition such as practice or study. Finally, a memory must be retrieved in order to be accessible at a later time. It is at this retrieval stage that memories are particularly susceptible to distortion. Once long-term memories are formed, they are not permanent, nor are they stored as intact units that can be simply retrieved and replayed. Rather, the information that forms various elements of a memory is stored as unique parcels in the brain region responsible for each type of memory, much like digital bits of information on a computer hard drive. For example, memory of a face is stored in the visual cortex, while the sound of a voice is stored in the auditory cortex. Thus, a memory of a person speaking, which includes both visual and auditory components, must be reconstructed from these different components each time it is recalled.

Distorted or false memories can result from a disturbance in any of the three processes outlined above, though it is primarily in the reconstruction phase of memory retrieval (that is, the assembly of memories from building blocks) that memory alterations occur. As a memory is reconstructed, all the related events and experiences since it was first encoded – including news sources, discussions, interviews, interrogations, or even self-reflection – can intrude with its retrieval. If not practised or rehearsed, with the passage of time information that was originally encoded can lose salience, and more recent information can intrude to alter its reconstruction.

The brain is highly plastic and capable of positive change with experience and learning; however, brain plasticity can also be negative, in the sense that unused information rapidly decays if it is not re-accessed and reused. Importantly, when individuals attempt to recall certain events, the newly reconstructed memory reflects what they believe they experienced. Individuals may insist that the most recent iteration of their memory is the correct version, even when confronted with contradictory evidence. Because individuals may not recognize factors influencing the encoding, consolidation, or retrieval processes, they can report distorted or false memories with high confidence. This apparent confidence in distorted or false memories is often interpreted as evidence of accuracy, as occurred in the case of Ronald Cotton.

To more fully understand the role of memory in legal processes, we must consider that there are multiple memory systems in the human brain. Memories of facts or events that can be consciously recalled and expressed verbally are uniquely supported by the declarative memory system. Declarative memories rely on a brain structure called the hippocampus, which is active during memory encoding and retrieval but not storage; hippocampal damage results in amnesia, mainly through the inability to encode information. The hippocampal declarative memory system has a unique role in reassembling information from across the brain to reconstruct memories, and it is this system – the hippocampal "mining" of other parts of the brain – that is susceptible to bias or distortion of memories during recall. In contrast, there is a different memory system, specialized for processing emotion, that is much less malleable. This memory system, mediated by an anatomically and functionally distinct brain structure, the amygdala, is crucial for the acquisition and expression of fear conditioning, through which we associate a particular stimulus with a frightening event. These two memory systems, hippocampal- and amygdala-based, do not operate independently. Rather, information from the amygdala is believed to enhance hippocampal-dependent memory with emotion. During the consolidation phase of memory formation, the amygdala modulates hippocampal memories through the actions of stress hormones. The net result is that events that elicit strong emotional responses, and are likely to be more important for survival, are also more likely to be well and more accurately remembered at a later time. These types of memories are thus less susceptible to distortion and, in some cases, impossible to forget. Indeed, the amygdala is thought to play a role in the repeated recall of highly traumatic memories, such as those associated with post-traumatic stress disorder.

The extent to which the different memory systems are involved in recording and recalling the events we experience will influence the susceptibility of any particular memory to alteration over time. It is difficult to know which memory systems are operating at any given time, but research clearly shows that eyewitnesses can develop distorted memories of the past. For example, merely suggesting to people that the perpetrator of a crime they witnessed had been holding a weapon may lead people to falsely recall the perpetrator holding a gun instead of a cell phone. Additionally, when eyewitnesses are asked to make an identification from a lineup, they may select the lineup member who most closely matches their memory of the perpetrator. If they select an innocent person, eyewitnesses may come to "remember" the selected individual as the perpetrator. Future choices are then guided by this distorted memory of the perpetrator's appearance. Importantly, individuals may become more confident in their distorted or false memories over time. This is particularly true if the memory is reinforced by repeated reports of similar details by other witnesses or by law enforcement affirming the identified person as the perpetrator. By the time a case is heard in court, a witness's distorted or false memory may be reported with very high confidence, potentially creating dire legal consequences for an innocent person. In fact, the Innocence Project, an organization that works to exonerate individuals (including Ronald Cotton) who have been wrongly convicted, reports that over 70 percent of their exoneration cases involve mistaken eyewitness identifications.

Individuals can also develop entirely false memories of putative autobiographical events. One can have a "memory" that is coherent, rich in detail, infused with emotion, and held with confidence, even when the event remembered never happened. Take, for example, reports by adult complainants of sexual abuse that occurred years earlier when they were children. In the majority of these cases, adults report accurate memories of abuse, and there should be no reason to doubt the validity of the accusations. In a small number of cases, however, the adults report having a completely repressed memory of the abuse and recovering it after extensive "memory recovery" efforts. The techniques used to recover memories can sometimes lead complainants to form false memories of past events.

The same cognitive processes that lead to false memory in complainants can lead to false memories in innocent suspects. Throughout an interrogation, investigators may use several manipulative tactics, including inviting the suspect to imagine how he or she might have committed the crime and asserting that there is no explanation for the evidence other than the suspect's guilt. These suggestive tactics lead some suspects not only to believe that they committed a crime but also to report specific details about the crime itself. Such details, supplied by the investigator, can provide compelling evidence of guilt, potentially leading to wrongful convictions.

The consequences of distorted and false memory vary widely. Falsely believing that you ate an orange for breakfast yesterday has little impact on your life. Conversely, distorted or false memories of a crime can lead to serious miscarriages of justice. The challenge for the court system is that the memory of a traumatic event (such as an assault), encoded by the amygdala, can be highly accurate in a general sense, even if some elements of the recalled memory (reconstructed in the hippocampus) are subject to potential distortion. Yet legal systems around the globe could not function without humans' memories of the past. Thus, a deeper understanding of the intricate interplay between memory systems is essential to ensuring the fair administration of justice by understanding how memory works. Luckily, there is substantial research and advocacy for improving legal practices to decrease the likelihood of contaminating the memories of both victims and suspects. For example, researchers are investigating the effectiveness of police interrogation practices designed to gather information in an unbiased fashion rather than a single focused effort to elicit confessions. Additionally, researchers who study eyewitness memory have developed lineup procedures to reduce erroneous identifications.

Given how heavily we rely on memory for recounting the past, it is encouraging that the legal field has made strides in embracing and integrating memory research into their procedures. And while progress has been made, researchers continue to study memory with the hope of providing legal professionals with best-practice recommendations for retrieving accurate memories. Ronald Cotton and Jennifer Thompson, who became close in the years following Cotton's exoneration, have contributed to this ongoing debate through joint public appearances and through the publication of their 2009 *New York Times* best-selling book, *Picking Cotton: Our Memoir of Injustice and Redemption*. As we seek to redress trauma imposed by historical and present-day injustices, we must exercise both empathy and care in considering the role played by our memories in understanding and communicating the past.

# DOCUMENTS OF DISSENT

*Laura Osorio Sunnucks, Gwyneira Isaac, and Diana Marsh*

ANTHROPOLOGY HAS BEEN CALLED a mirror for humanity, a reflexive practice that can help us understand ourselves through the study of others. As storehouses of material culture and written records, museums and archives can contribute to this process, transmitting knowledge from one place and time to another and providing a lens through which we may come to know ourselves better. Objects of knowledge in these institutions are used to communicate the past to the present and the present into the future. In the case of anthropological collections, the knowledge is cultural: it relates to the collective identities and memories of communities.

The role of museums as the guardians of memory artifacts is not, however, without problems. The practices of colonialism and cultural appropriation have made engagement with these institutions and their collections challenging. As Linda Tuhiwai Smith, of the New Zealand Ngāti Awa and Ngāti Porou peoples, states, the "collective memory of imperialism has been perpetuated through the ways in which knowledge about Indigenous peoples was collected, classified and represented in various ways back to the West, and then through the eyes of the West, back to those who have been colonized."[1] Addressing this problem requires museums to critically appraise their acquisition processes and representational choices, past and present, and, where possible, to return museum collections to marginalized source communities. American anthropologist Ruth Benedict suggests that this alone is not enough, because the hegemony of the colonial world is reasserted through the very study of the colonized world; to see ourselves in others can be to impose our own perspectives on those whose artifacts we hold in our collections. How, then, can museum documents and material culture retain utility as legitimate guardians of collective memory? Is there a role for museums in a world that strives to be postcolonial?

Memory evokes a meaning distinct from the idea of history. Whereas history suggests a document, text, or object that carries a version of contemporary truth, memory implies a partial narrative that may warp or diminish over time. Human memory is subjective; it cannot be stored simply. Therefore, it may be seen as a questionable tool in a world where objectivity and truth are privileged. Moreover, since cultures and societies don't think, feel, or remember – people do – can there be such things as collective or common memories? Surely, one might argue, history (constructed through texts and material objects) offers more reliability.

Memory does allow us, however, to go beyond a national or dominant narrative to challenge the official written records that become history. In this way, popular memories, and the objects associated with them, expand the historical archive, bearing witness to meaningful experiences. This is particularly important for communities whose past and present cultures are oppressed and whose "truths" are not represented in mainstream culture. Historical artifacts and texts can help reconnect disenfranchised groups, including Indigenous communities, with lost or stolen forms of knowledge. In theory, museums, archives, and their collections can offer this help. In practice, however, the objects collected in museums may not be readily available to those from whose communities they were derived.

For more than two hundred years, for example many Indigenous peoples did not have access to the sources of their cultural and linguistic knowledge stored in museums. Beginning with Thomas Jefferson, the study of Indigenous languages in North America was an extractive process, taking knowledge from Indigenous peoples and transposing it into documents stored away in places such as the American Philosophical Society, in Philadelphia. In some cases, these archives would come to serve an important role in preserving languages that had lost their last fluent speakers. That said, these repositories were only useful to the extent that they remained open and accessible to all. Until recently, only scholars with the right research credentials could access archival records in the American Philosophical Society, effectively marginalizing Indigenous people who might wish to consult them. Now that these barriers have been removed, important early records and documentation are supporting cultural resurgence and language revitalization in the communities from whose ancestors these materials were collected.

The digital era offers a new promise of access, creating opportunities to virtually connect communities with objects of knowledge and memory that were previously lost or appropriated.[2] In this way, archives can play a critical role in the recovery of a collective history, preserving items key to the memories of a people. Indigenous scholars and knowledge keepers are increasingly using archival records in North America to regain sovereignty over the memory of languages, cultures, and places. Sometimes, these records replicate what communities already remember through oral tradition, but the artifacts nevertheless have immense power when used for community purposes.

In addition to enabling projects that strengthen a community's relationship with their histories and heritage, museums and archives can also be spaces that stimulate discussion on contemporary Indigenous and minority issues. Museums can, for example, keep the memory of contemporary atrocities against Indigenous or minority communities alive. Consider the following recent example. On September 26, 2014, forty-three mostly Indigenous, male students from the Escuela Normal Rural Raúl Isidro Burgos in Ayotzinapa (a rural school for Nahuatl-speaking teachers in Guerrero, Mexico) were forcibly "disappeared" following a clash with municipal police in the nearby city of Iguala. The authorities eventually reported that these students had died at the hands of a local drug cartel. It was later alleged that the police had handed the students over to the cartel under orders from the mayor of Ayotzinapa and his wife, who have been shown to have close ties to local cartels. Despite significant media attention and an ongoing investigation by an international research agency, the full particulars of the disappearances have never come to light. The government has falsified evidence and subverted the justice system to conceal the extent of its involvement. Three years on, the families of the victims continued to press the Mexican government for transparency and a fair investigation of these crimes.[3]

Amid the subversion of justice and concealment of truth, certain facts resound. Guerrero is one of Mexico's poorest and most federally neglected states, a place where there is little distinction between the government and organized-crime units. In addition, the Ayotzinapa disappearances, while covered both nationally and internationally in the press, are a clear example of the way that marginal and rural (and, in this case, predominantly Indigenous) peoples in Mexico continue to be made victims of state-sponsored violence and excluded from dominant societal democratic structures.

Shortly after a small part of the students' remains were found, researcher Juan Manuel Sandoval Palacios, from the Instituto Nacional de Antropologia e Historia (INAH), Mexico's governmental heritage institution, and his son, Diego Sandoval Ávila, created the Ayotzinapa Codex, a pictorial manuscript that uses iconography from the colonial era to denounce neocolonial abuses in contemporary Mexico. The work likens the forty-three disappeared students to warriors captured by the Aztecs in their pre-conquest empire-building wars. It also casts the current government as the ideological descendant of the sixteenth-century colonial administration in Mesoamerica. Mexico's contemporary Indigenous peoples are associated with the subjects of historical New Spain, and the neoliberal forces in contemporary Mexico are concomitantly associated with the racist ideologies of the Spanish colonizers. In its repudiation of both Spanish and Aztec colonial methods of

Fragment of the Ayotzinapa Codex, by Juan Manuel Sandoval Palacios and Diego Sandoval Ávila.

subjugation, the codex critiques dominant modes of governance, irrespective of culture and society.

The work has been periodically exhibited in the public entrance hall of Mexico's largest and most frequented public museum, the Museo Nacional de Antropología, where families of the victims went before an audience of Mexican and international visitors to fundraise for their cause. Measuring over eight metres in length, the codex has been used as a powerful visual aid at political demonstrations in Mexico's capital city, condemning the Ayotzinapa disappearances. In preserving and exhibiting this work, a museum might play a central role in sustaining the memory of an atrocity committed against marginalized Indigenous people.

The artists behind this work, and the supporting INAH syndicate, consider the work to belong to the families of the forty-three students from Ayotzinapa, as a testament to their struggle for restitution of their legal rights. This view of its ownership could have provoked resistance to selling the work or allowing it to be displayed outside of Mexico. The artists were recently approached to sell the work to the Museum of Anthropology at the University of British Columbia. The museum felt that the document could be mobilized as an important memory object in their Latin America collection. After initially rejecting the sale, the artists consulted with the students' families and agreed that the proceeds of such a sale might be used to advance their cause. Furthermore, the families of the victims believed that the inclusion of the codex in an international museum would provide significant exposure outside of Mexico for their work.

There is much to learn from this acquisition process. Works such as the Ayotzinapa Codex preserve memories of human rights violations that too often lack representation in heritage institutions. In its use of pre-Hispanic iconography and form, the work prompts its audience to explore questions surrounding the legacies of colonial history. We can also see how the piece encourages us to reflect on the ways in which the past can be appropriated and represented in the development of contemporary societies. The process of acquiring this object for a Canadian museum may also represent a self-reflexive and critical moment in the development of museum collections. Had it not been for the international lobbying efforts of the families of the forty-three students, the artists might not have considered the sale of this important work, particularly to a non-Mexican institution. Complex political and social contexts underlie the acquisition of all museum collections, a fact that (aside from concerns over colonial processes in the shaping and legitimizing of collections) is often overlooked. The story of the codex's acquisition is a prompt to museum visitors to consider the multiple ways in which objects come to form part of collections, lending another layer of complexity to the relationships that museums aspire to create with the public.

As communities use collections to reassert their identities and their political agency, museums and archives can become sites of return and remembering for many peoples. Projects that involve the virtual reunification of communities and archives or that promote the culturally sensitive care of (and physical access to) meaningful collections ensure that previously excluded cultural stakeholders can use collections to activate memories. This kind of work must be mindful of the need to rectify the damage caused by the museum's or archive's colonial legacy. Objects such as the Ayotzinapa Codex and the process by which they are included in anthropological museum collections broaden our perspective. Museums and archives are not neutral spaces. Rather, their struggle for relevance requires self-reflection and, potentially, taking a committed political position. It is imperative that museums not only store artifacts but also actively facilitate the memories and rebuild the communities that are central to their social value.

NOTES

[1] Linda Tuhiwai Smith, *Decolonizing Methodologies: Research and Indigenous Peoples* (London: Zed Books, 1999), 2.

[2] Anne J. Gilliland, "Permeable Binaries, Societal Grand Challenges, and the Roles of the Twenty-First-Century Archival and Recordkeeping Profession," *Artifacts* (December 2015): 12–30.

[3] Ryan Devereaux, "Three Years after 43 Students Disappeared in Mexico, a New Visualization Reveals the Cracks in the Government's Story," September 7, 2017, *The Intercept*, https://theintercept.com/2017/09/07/three-years-after-43-students-disappeared-in-mexico-a-new-visualization-reveals-the-cracks-in-the-governments-story/.

# ANTHEMS

*Ian Williams*

IN THIS LYRIC ESSAY, I perform a simple act of memory. I attempt to recall the words of the national anthems of the countries where I lived and worked. How long does patriotic information last? The retrieval was done without any aids – just me burrowing through the accumulated sediment of years.

The essay imitates the structure of my memories: at the centre sits a faded nucleus of truth, superimposed over that is the recollection, and branching from that is a network of associations.

## LIVED

The anthem begins with the Jehovah's Witness kid leaving the room and leaning against the lockers with her arms folded.

O Canada —
O Canada!

Our home and native land
Our home and native land!

True patriot love in all our sons' command.
True patriot love in all of us command.

With glowing hearts we see thee rise
With glowing hearts we see thee rise,

There's a brass interlude of grand descending blasts that lead to the word God.

the true North strong and free.
The True North strong and free!

From far and wide, O Canada,
From far and wide,

we stand on guard for thee.
O Canada, we stand on guard for thee.

The God of this line never seemed like the God of the church I went to. He had no real plan for keeping the land glorious and free whereas the God from church had clear ideas about fire, water, strong towers, rocks, refuge, and fortresses.

God keep our land glorious and free.
God keep our land glorious and free!

~~Protegera nos foyeyers et nos droits.~~ *Ah*.
O Canada, we stand on guard for thee.

O Canada, we stand on guard for thee.
O Canada, we stand on guard for thee.

The older I got, the more this part of the anthem seemed like the lyricist had run out of material. Three times! We stand on guard for thee.

Don't nobody want to invade Canada.

Maybe Ottawa could commission Atwood for a rewrite.

I never heard an exclamation there. There's nothing in the introduction to suggest *O Canada!* [Flap a wrist] *You so crazy!*

As a child, I thought this meant that the sons of Canada had the power to command love instead of love having the power to inspire our sons.

Nope. *Thy*.

Childhood mnemonic: Imagine the sun rising up, up, up to the true North like a helium balloon.

CBC: Senate passes bill to make "O Canada" lyrics gender neutral.

Whenever I was paying attention, I imagined literal guards standing in front of Buckingham Palace wearing Mountie gear crossed with British Royal Guard gear. They also held bayonets upright, with the swords part extended and leaning against their shoulders.

A stutter of memory. First came the music, then the words in French, then the words in English.

Sometimes an inspired teacher would precariously climb the anthem to the high note at the end.

The Jehovah's Witness kid returns with her head down. The principal says, Good morning. It's now time for today's announcements.

# WORKED

The only time I needed to sing this anthem was at a convocation ceremony in Massachusetts. Everyone turned towards the flag, which we never did in Canada, and placed their hands over their hearts. I thought I knew the words. It turned out I mostly knew the melody.

Oh-oh say can you see, by the dawn's early light.
O say can you see, by the dawn's early light,

What so proudly we gave dada da da da da
What so proudly we hail'd at the twilight's last gleaming,

And the rocket's read glare!
Whose broad stripes and bright stars through the perilous fight

While the ramparts bright stars dada da da da da da
O'er the ramparts we watch'd were so gallantly streaming?

And the rocket's red glare, the bombs bursting in air,
And the rocket's red glare, the bombs bursting in air,

da da da da night but the da was still there,
Gave proof through the night that our flag was still there,

O say does that star-spangled banner yet wave
O say does that star-spangled banner yet wave

for the la-a-and of the free and the home of the brave.
O'er the land of the free and the home of the brave?

*freeeee*

If this were an action song, one would bend one's arms at ninety degrees and march in place here.

I never actually considered that the anthem could be sung by anyone who was not a celebrity standing on a platform in a football field. She is wearing a sparkly pantsuit, holding her ear with one hand and measuring the pitch of her runs with the other.

The anthem should get more radio play. The invisible supermarket DJ could mix it in between "Torn," "Kiss from a Rose," and "Bailamos."

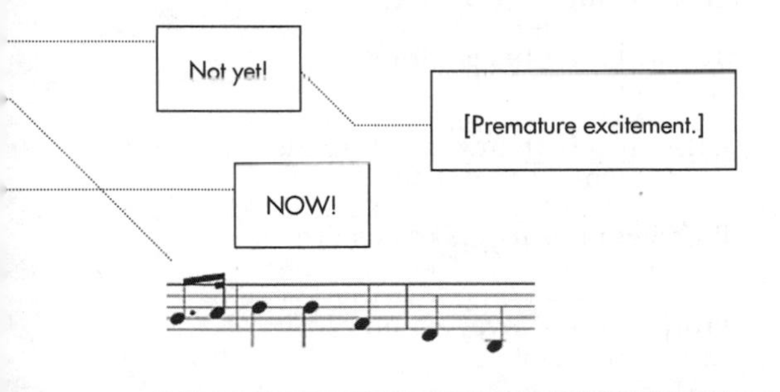


The football players are kneeling but not praying.

In memory, the American anthem always ends with applause, cheering, and something like fighter jets shooting across the sky.

Oh, it's a question?

# AM

The French immersion kids were particularly strong anthem singers. They sang it as effortlessly as the rest of us sang the *Growing Pains* theme song.

*Pron.* "cat on bras"

Because of *liaison*, which was the most sophisticated word I knew for a while, until it was surpassed by *antagonism* in an epic Grade 6 speech on the fall of the Berlin Wall.

O Canada, terre de nos aieux.
Ô Canada! Terre de nos aïeux,

Ton f ont essaie des fleurons glorieux.
Ton front est ceint de fleurons glorieux!

Car ton bras est Pompéi.
Car ton bras sait porter l'épée,

Il s'est porte le froid.
Il sait porter la croix!

Ton histoire est une Pompéi.
Ton histoire est une épopée

Tes brillia-ah-nts exploits.
Des plus brillants exploits.

Laisse nos fritteurs! Ton font essaie.
Et ta valeur, de foi trempée

Protègera nos foyers et nos droits.
Protégera nos foyers et nos droits.

Protègera nos foyers et nos droits.
Protégera nos foyers et nos droits.

I still imagine bright explosions on Canada Day night, although I know *les exploits* have nothing to do with explosions.

*Google trans.*
Your forehead is girded with glorious jewels.

There's a great line by Atwood —

No, it's by Plath, in "Tulips":
I am nothing. I have nothing to do with explosions.

Just checked. The actual line is:
I am nobody; I have nothing to do with explosions.

Because we learned the sounds of the words before we learned the vocabulary, *terre de nos aïeux* sounded to me like *terre de nos errors,* "land of our errors." Then it sounded like the nearest French word I knew at the time, *oiseau*. Land of the birds. How very much Canadians love geese, I thought, to place them in the anthem and on the dollar and on the lakes of city parks.

A homophonic slip-up. Close.

Not actually a word. *Etym.*

*valeur*
˅
*friture*
˅
*fritter*
˅
*fritteur*

*Pron.* "glory-yeah"

The French *r* is difficult for English speakers. To pronounce it, I've been told to say *r* and *h* at the same time, to gargle *car,* to scrape the back of my throat for tonsil stones.

The gender of French words has nothing to do with gender in real life, yet I assumed that *croix,* with its terminal *x* and its dying Christ, would be masculine.

Alarmingly, Pompeii is referenced *twice* in my recollection of the *Canadian* national anthem.

The first new car my mother owned in Canada was a grey Mazda Protégé. It had a congenital rust problem.

*Trans.* [childhood]: Protect our tile foyers and our right turns.

Encore.

# IN DEFENCE OF FORGETTING

*Shannon Walsh*

IT IS A CULTURAL AXIOM that we must remember the past or be condemned to repeat it. But as experience has shown us, at times there are things best left forgotten. The surety of vindication is reserved for the privileged. Most of us are not so lucky. To move on, to move forward, to keep going, to persevere – to do these things, forgetting (or at least not keeping something close in mind) has its uses. In a world of violence and trauma, forgetting is as elemental to human action and human life as is remembering. What if it is survivors who best know the benefits of forgetting and the dangers of collective remembering?

The role of memory in our society rose to collective consciousness as the #MeToo movement rocked centres of power, from politics to entertainment and business. As women around the world began to share stories of sexual harassment, violence, and abuse, the full extent to which women have had to suppress their experiences came clearly into view. For many women, the process of dredging up the past has been difficult, even if it has also been encouraging to feel part of a collective moment of change. Nonetheless, questions linger: Who does remembering serve, in this moment? After all, for the most part, women have whispered among ourselves about such things since girlhood. Certainly, speaking out publicly and being heard, feeling united with other women who have gone through the same thing, should not be undervalued. Some women feel hopeful in telling their stories and in initiating processes that may result in some kind of justice. Yet, for others, such pathways are more muddled. Moments that are too hard to talk about, too ambiguous to lay claims on, too dangerous to risk speaking out loud, too fresh and too painful: these memories are sometimes best left dormant. There is a discomfort expressed by many women, an unease at the way women's stories have been repurposed by men attempting to exonerate themselves or wipe the slate clean (such as under the hashtags #IdidIt #ItWasMe). Other women may also not be ready to move on, to accept apology, to settle accounts in the midst of ongoing misogyny and violence against women every day. Surfacing a lifetime of memories of harassment might not be useful for many women; in fact, reviving it all is often further traumatizing. Does this collective remembering serve perpetrators more than it serves survivors? What are we to do with all these memories, hidden and now revealed?

There are many ways to think about the usefulness of forgetting. One way is to consider how trauma produces memory suppression. One of the symptoms of post-traumatic stress disorder, or PTSD, is avoidance, which includes the inability to remember key aspects of the trauma. Avoidance is an elemental way to cope with something extremely painful; the mind suspends normal operations in order to survive. In many ways, sexual assault survivors have been experiencing a collective trauma and a collective avoidance of the memories of harassment and violence that have been part of so many lives. What does it mean for millions of survivors to decide to remember out loud, together? What does it mean when such remembering is not a choice made by survivors?

Episodic memory, the ability to remember a certain moment or event, is fundamental to understanding ourselves, but forgetting is just as fundamental. Imagine living entirely without forgetting. Ruminating on every detail, unable to let slights, injustices, or heartbreaks fade, stuck in a spiral of images and emotions. Forgetting can be useful. Yet forgetting is often seen as a disability, an impediment to enlightened thinking, a flaw to be remedied. For a rare few, remembering everything is not just a philosophic question but an everyday reality. A rare neurological disorder causes a small percentage of the population to vividly remember everything that has happened in their lives. The condition is called hyperthymesia syndrome – derived from the Greek *hyper* (excessive) and *thymesis* (remembering) – and is also known as highly superior autobiographical memory, or HSAM. People with HSAM vividly recall almost every day of their lives, including both significant and mundane events, with as much detail as most of us remember events that happened yesterday. These mostly autobiographical memories are involuntary; they are stored without any effort and arise automatically. Not being able to forget is its own kind of punishment. Action itself relies on the ability to forgot, or at least suspend knowledge of, all of the potential consequences.

If human action is, in part, contingent on forgetting, what can it tell us about moments of collective remembering? Who does such remembering serve? Historical and cultural memories are often political, an attempt to impose particular narratives on the present and to justify certain present-day positions. The journalist David Rieff claims it is magical thinking to imagine that memorializing genocides, for example, will help to halt future ones, as evidenced by Rwanda, Myanmar, and Cambodia.[1] In these overtly populist times, memory may in fact promote violence and thwart forward motion. Nationhood is a complicated and often violent act of collective memory, one quite dangerous in its omissions. Just think of President Trump's "Make America Great Again" slogan, which many hear as an invocation of a white supremacist past or of an idealized bygone era that never existed at all. Although episodic memories do a great deal to help us understand ourselves and our place in the world, collective memory is often a way to narrow the narrative, to homogenize different stories to create a single master accounting of the past. Who shapes what is included or excluded in the master narrative is often a question of who holds, or retains, power. The enduring effects, and affects, of colonialism and its continued domination over the stories that are told about race, place, and nation are stark reminders of that power.

The act of remembering, rewriting, and repopulating history with (and by) those peoples who have been forgotten or left out has been an important part of redress over the last half century. Reconciliation with First Nations peoples is one example of the current preoccupation with remembrance as redemption. I remember speaking to a Chipewyan Cree Elder while making a documentary in northern Alberta and asking him to tell me about the environmental changes he had seen on the land as a trapper. He was shaken and saddened. Only a few months before we spoke, he had been asked to recount his experiences at residential school, where he and thousands of other Indigenous children had been forced to attend and where many had endured emotional and physical abuse. Each year he suffered in residential school was fixed to a dollar amount for compensation under the Indian Residential Schools Settlement Agreement. Many people in his community were living in extreme poverty, in the biting cold winter. Their extracted stories got them a few hundred dollars but left them traumatized all over again. The Elder described the process as opening a can of worms that could no longer be closed. Afterwards, the scribes and government agents left, their job done, compensation delivered. The Elder said he was struggling to cope with the emotional toll of reliving the violence he had endured as a child while also struggling, as an adult, with the destruction of his peoples' lands and territories. Too much at once. Forgetting is not an option, but perhaps there should be a right to refuse to remember? Alia Somani writes about Canada's fixation on the apology and its inherent paradox. She explains that while the apology attempts to find redemption and seal the past, it also opens up the trauma and "has the potential to resuscitate memories of the past, even as it attempts to suppress them."[2] Remembering is essential, for some unavoidable, but there are times when its public display serves the most powerful, a cover to hide behind and to assuage guilt.

While remembering past injustices done to Indigenous peoples in Canada is crucial to any reckoning with what Canada was and can be, remembering has its dangers. Collective remembering often denies survivors the right to forget because their individual stories of trauma are needed to initiate liberal reconciliatory processes. These processes often focus on fixing or healing survivors' trauma rather than on fixing the conditions and structures that produced and enabled the trauma and that continue to reproduce it. In the aftermath of collective remembering, an apology is offered and forgiveness is deemed to have been achieved. But individual victims have not truly been part of the process of forgiving and, worse, the conditions that produced the harms in the first place have not been addressed. Instead, reconciliation with Indigenous peoples in Canada is often an appeasement for white-settler guilt. Settler communities undeniably benefit from reckoning with the past in a manner that is predetermined towards forgiveness and the continuation of the status quo. When you consider the injustices done to Indigenous peoples – treaties broken, children stolen, land pilfered and occupied, thousands of women disappeared, murdered, or raped, not to mention cultural and physical genocide – a political process built from the outset to engender forgiveness through apology is generous to those at fault and to those who have benefitted from a system of racial supremacy. As Gwitchin filmmaker and writer @kwene_dru succinctly posted on Twitter: "Reconciliation is only for non-natives to absolve themselves of a collective guilt, and all it has done is further traumatize indigenous people."[3]

The demand for collective remembering can be used as an assertion of power, forcing the acceptance of an apology without the consent of individual survivors, relegating injustice to a past that survivors must now "get over." In *Red Skin, White Masks: Rejecting the Colonial Politics of Recognition*, Glen Coultard, a Yellowknives Dene scholar, writes that "state-sanctioned approaches to reconciliation must ideologically manufacture [a transition from an authoritarian past to a democratic present] by allocating the abuses of settler colonization to the dustbins of history." Such an approach denies any reckoning with the present; it silences contemporary complaints about ongoing injustices. As he writes, "If there is no colonial present ... but only a colonial past that continues to have adverse effects on Indigenous people and communities, then the federal government need not undertake the actions required to transform the current institutions and social relationships that have been shown to produce the suffering we currently see reverberating at pandemic levels within and across Indigenous communities today."[4]

The focus on collective remembering as a means to forgive and forget disavows survivors their very understandable resentment and anger for the ongoing harms of a settler-colonial present. These ongoing affective responses, Coultard argues, are both entirely appropriate to the ongoing "structural and symbolic violence" Indigenous peoples face, and such resentment and anger can be key tools for creating a rupture from colonial subjugation.[5] Survivors should have the right to choose when to remember, how such remembering is framed, and the terms on which the remembering is done. Collective remembering can at times be silencing rather than empowering for survivors.

The space between remembering and forgetting is a place of pain for many people in Canada. It is a space where antagonisms are sometimes buried and, too, where survivors find the strength to choose to either move on or hold on. Those who need to remember and to act on their complicity in oppression are often not called to do so in meaningful ways. Instead, the act of remembering is laid at the feet of survivors. Survivors may find use in forgetting, or in refusing to remember on the oppressor's terms. Collective remembering must be attentive not only to the power dynamics of who is allowed to forget and who is forced to remember but also to what is remembered and what is forgotten. It is important to beware of a liberal reconciliation politics meant to absolve issues of misogyny and settler colonialism that are far from over. When unearthing a trauma, the question that should always be asked is, Who is truly served? It may seem obvious that bringing to light collective memories of sexual harassment and colonial injustice is the only way to reconcile the past, yet it may well be that the risk is less in forgetting but in allowing those in power to use collective remembering to reproduce the conditions that created trauma in the first place.

NOTES

1 David Reiff, *In Praise of Forgetting: Historical Memory and Its Ironies* (New Haven: Yale University Press, 2016).

2 Alia Somani, "The Apology and Its Aftermath: National Atonement or the Management of Minorities?" *Postcolonial Text* 6, 1 (2011): 7.

3 Kwene_Dru (@kwene_dru), Twitter, February 11, 2018, 12:18 p.m., https://twitter.com/kwene_dru/status/962782849588912128.

4 Glen Coultard, *Red Skins, White Masks: Rejecting the Colonial Politics of Recognition* (Minneapolis: University of Minnesota Press, 2014), 121.

5 Ibid., 115.

# MONUMENTS IN STONE AND COLOUR

*Hanna Smyth*

THE DOMINANT IMAGE of the First World War is of white soldiers fighting in trenches along the Western Front. War poetry by white men is taught in schools across North America and Europe, and the deeds of white men are remembered through their medals in museums. During anniversaries of the conflict, people holding photos and telling stories of their white First World War ancestors feature heavily in the media.

While important, this picture does not represent the whole story. The First World War had many fronts, all over the world, and the conflict required several million people to travel large distances to foreign countries as part of their war service. Whose mental picture of the First World War includes a man from the Chinese Labour Corps who died after the war while digging graves in France? or the Japanese Canadian soldier who returned home to a country that still didn't treat him as equal? What of the prison guard from the British West Indies Regiment who served in Egypt? or the Maliseet man who served as a scout on the Belgian front line? Women of colour also participated significantly in the war, on the home fronts and, more rarely, in medical services, yet they are far too absent from remembrance. As we move beyond the centenary of the First World War, we need to continue challenging and diversifying our understanding of who was part of this war and who qualifies to appear in its remembrance.

Because of the racial prejudices of the time, many men of colour were not permitted to fight in the war and were instead used for labour essential to the war effort. Although our conceptions of the war are dominated by images of soldiers holed up in trenches, these soldiers relied on a large and less visible (at least to outsiders) support network. The soldiers needed people to lay railway tracks for transport; they needed people to cook, cut down trees, and dig communication trenches. The various labour corps who engaged in such work were not permitted to have weapons and were not expected to fight. Yet they still undertook dangerous and difficult work – often very close to the fighting – and many of them were killed.

Three particularly significant contributors of men for First World War labour were South Africa, India, and China. The South African Native Labour Corps comprised more than twenty thousand people. Among them were respected warriors and leaders, yet these men were not allowed to carry weapons or mix with white communities. The force was first sent to German South West Africa and East Africa, and they arrived in northern France by 1917. They toiled under intolerable and dangerous conditions and played an important role in the eventual allied victory. Many were injured and killed.

As part of the British Empire, India was automatically pulled into the conflict once Britain declared war. The country raised 1.5 million men for the First World War: the largest volunteer army in history, comprising over 800,000 soldiers and over 500,000 noncombatant labourers. The Indians who served in the First World War – including Muslims, Sikhs, and Hindus – were from diverse backgrounds and places, including what we now know as the modern state of India and territories that would later become Pakistan, Bangladesh, and Sri Lanka.

China's labour corps had at least 140,000 men. These men worked as labourers for both the French and British armies, in factories and shipyards, handling ammunitions, repairing roads, and much more. China had agreed to provide men as labour for the Allies on the understanding that it would have greater influence at the Paris Peace Conference at the conclusion of the conflict. Most of the Chinese Labour Corps served in France, and thousands of Chinese men were secretly transported there in locked trains and in deadly conditions across the entire length of Canada. Technically, this remains the largest mass migration in Canadian history, though it is not widely taught or known.

People of colour in America and the white Dominions – including Indigenous peoples (the Māori of New Zealand, for example) and minority groups such as African Americans and Japanese Canadians – also contributed in both combative and labour roles. Although their pre- and postwar rights differed across groups and between Dominions, all shared the unfair predicament of participating in the war effort of countries that did not view them as equal citizens and that denied them equal rights both before and after the war. Hopes that war participation would enhance equality rights – a hope explicitly stated by many who enlisted – were slow to materialize.

In the absence of any surviving First World War veterans, we have to turn to other sources of information to understand who these people were and how we might best remember them. Material culture – tangible things, specifically memorials and cemeteries – are a crucial source of data. The material culture of First World War remembrance, in the form of stone memorials and cemeteries constructed worldwide, marks not only sites of remembrance but also sites of identity. These locations (often on battlefields and far from the homelands of the dead they commemorate) are world stages upon which nations seek to create historical narratives of their wartime sacrifices under one another's watchful eyes. Stone memorials "perform" in the sense that they were created with audiences in mind: the international mourners, tourists, and politicians who would see and judge them.

The choices made concerning the representation (or lack thereof) of people of colour in the material culture of remembrance convey significant messages not only about the dead but also about the living who made these decisions. At the centenary of the conflict, examining how these sites are used in public and private remembrance rituals, and how their meanings have changed, is crucial for understanding both the dead and the survivors of the First World War – all of whom are now gone.

First World War memorials and cemeteries are dynamic places. They may be made of stone, but they are always changing. Following the war, it took almost twenty years for the memorials and cemeteries to be built, and they were in constant flux: designs and architects changed, new bodies of soldiers were discovered, names were added to and removed from the lists carved onto memorials, and names on headstones were changed when people discovered that soldiers had been serving under false names. These sites are still dynamic today – substantial work goes into maintaining and restoring them, and memorials are alive with visitors, especially during the centenary. Additional bodies are also still being located, new memorials are being erected, and people continue to ascribe new meanings to those already there.

Memorials thus do not spring out of the earth. They are designed and derived from the results of hundreds of decisions, big and small, made by many different people. These decisions impact what we do and don't see when we visit a memorial, and they affect how we understand what we see. Memorials and cemeteries are also not neutral: they are designed to make the visitor think certain things and feel certain ways. The people and institutions making these decisions have considerable power to shape our memories and interpretations of past events. They also have a great responsibility to honour the dead in a manner that is acceptable to a diverse "audience" with potentially conflicting sensibilities.

The designers of the national memorial for India at Neuve Chapelle in France had to navigate complex emotional and cultural terrain. Britain made the decision to have a single memorial for all Indian forces over the protests of the Indian government, which argued for separate memorials for the men of each religion. The conception of a unitary memorial posed the difficult task of designing a structure to commemorate Hindu, Muslim, Sikh, and other Indian soldiers in a way that would satisfy all religions yet privilege none. This memorial honours the missing, listing the names of men whose bodies were never found. For those bodies that were found, complex and contradictory religious beliefs regarding treatment of the dead, particularly between Hinduism and Islam, posed substantial challenges to the provision of appropriate burial and commemoration. In some instances, Hindu soldiers were cremated according to traditional practices, either on the battlefields or after they had died in hospital. For example, some of the Indian soldiers who died in the United Kingdom in the Royal Pavilion at Brighton were cremated nearby on the South Downs. Shortly after the war, in 1921, the Brighton Chattri memorial was constructed at the site of the cremations.

Place offers an added layer of meaning to memorials. For those memorials erected on the site of historic battles, the specific location of a commemorative site carries implicit and coded messages about which groups and events deserve pride of place in national remembrance. For example, South Africa's national memorial at Delville Wood, France, was ostensibly for all South Africans who had served. But the choice of location (at a site where only white soldiers had fought) and the sculpture itself (twins representing the two "white races" of South Africa) suggest that this was conceived as a largely white site of remembrance. In 2016, Jacob Zuma, the president of South Africa, unveiled a new addition to the memorial, honouring South Africans of all races. The new addition is a wall of names, listed in alphabetical order, that identifies all of the South Africans who died, irrespective of colour.

Memorials can also be erected far from the sites of actual combat, providing an opportunity for remembrance in the home countries of those who travelled great distances to fight wars on foreign soil. A lasting effect of imperialism is that such domestic commemoration can occur either in the country itself or in the colonial motherland. For example, the New Zealand War Memorial in central London was designed to take the form of markers at Māori ancestral sites. In 2016, Canada unveiled a statue to honour Francis Pegahmagabow, a decorated First Nations sniper, in his home town of Parry Sound, Ontario. Both these memorials – one in Great Britain and the other in Canada – act as proximal sites to commemorate the actions of individuals who lost their lives fighting in another country in the service of the British Empire.

As we seek to create a more nuanced understanding of who belongs in First World War memorialization, we must open our eyes to hidden presences and, equally important, to absences. When we visit war memorials or battlefield cemeteries, or when we engage with other forms of First World War remembrance, we would do well to ask ourselves what is missing and what is going unacknowledged. Whose perspective, or voice, or experience isn't represented in stone, and why? As the centenary closes, we must continue to ask such questions, at spaces that commemorate not only the First World War but also other historical and contemporary conflicts. In this way, we will more fully acknowledge the memory of all those who have lost their lives in the service of their countries.

# MICROCOSMOS

*Steven J. Hallam*

THE ORIGIN AND DIVERSITY of life on Earth has inspired countless stories of creation, cataclysm, and rebirth since human beings first began to wonder. Our early ancestors likely directed much of their curiosity outwards, looking for traces of the origins of life in the visible world and in celestial objects. This began to change in the seventeenth century, when Anton van Leeuwenhoek gazed through the first microscope, revealing an enigmatic and previously invisible microbial world, the microcosmos. Since that time, we have learned how to read the evolutionary memory of Earth as it is encoded in this invisible world. Through careful study of conserved gene sequences and ancestral protein function, the emergence of cells and distributed metabolic networks, we can see how microbial life has both shaped and responded to changes in the Earth system over evolutionary time. Here, we may also find the inspiration and technical savvy needed to solve some of the most pressing environmental problems of our time.

At its most fundamental level, the replication and expression of biological information is shared by all living things. Nucleic acids (DNA and RNA molecules) encode genetic blueprints for life, which are translated into proteins that build cellular structures and catalyze the biological reactions that drive cellular metabolism. This flow of biological information can be traced back to primordial chemical reactions in early Earth history and the emergence of self-replicating genes. Enclosure of these genes in lipid membranes gave rise to the first cells, which, in turn, became information-storage and -processing units in an evolving ecological network that continues to this day. By studying variations in the genetic information contained in organisms, we can reconstruct evolutionary relationships between extant (and extinct) life on Earth and understand how this life has both shaped and responded to changes in the planet's physical and chemical environment.

Take, for example, the genetic information encoded in ribosomes. These subcellular protein factories play a critical role in translating information from genes into the proteins that drive cellular metabolism. The biochemical activity encoded in the RNA structure of the ribosomal machinery can be traced back to a moment in deep evolutionary time when the first complex proteins were formed from simple building blocks (free amino acids) found in the early Earth environment. All cellular organisms contain ribosomes, and these structures have remained relatively constant over billions of years of evolution. However, over time, small changes in ribosomal RNA gene sequences (mutations) have randomly occurred and subsequently been inherited within organismal lineages. By comparing ribosomal RNA gene sequences from different organisms, we can place the organisms onto the proverbial Tree of Life, with the most closely related species sharing the most similar pattern of accumulated mutations as a shared evolutionary memory.

Memory is also written into the structure and function of the proteins produced by ribosomes. Consider the protein Ribulose-1,5-bisphosphate carboxylase/oxygenase (RuBisCO), which catalyzes a crucial step in photosynthesis, converting carbon dioxide and water into oxygen and sugar. RuBisCO is present in the vast majority of photosynthetic organisms – from cyanobacteria and moss to maple trees and water lilies – and it is directly responsible for most of the organic carbon naturally produced each year (about 100 billion tons globally). Despite its central role in Earth's carbon cycle, RuBisCO is an inefficient enzyme with a low binding affinity for carbon dioxide and an extremely slow reaction rate (it catalyzes about ten reactions per second, as compared to ten thousand to one hundred thousand reactions per second for other enzymes). The reasons for this conundrum become clearer when we consider the physical and chemical environment in which RuBisCO first evolved.

Approximately 3 billion years ago, a group of microorganisms (ancestors of the blue-green cyanobacteria) innovated a biochemical pathway that could separate water into oxygen and hydrogen, with the hydrogen used as a source of electrons needed for photosynthesis. Prior to the advent of these water-splitting reactions, Earth's atmosphere was rich in carbon dioxide and devoid of oxygen. Over time, free oxygen accumulated, altering the surrounding environment. The presence of oxygen posed a significant challenge for photosynthetic organisms because oxygen can react with RubisCO and decrease the enzyme's ability to bind carbon dioxide and produce sugars. As oxygen continued to increase in the early atmosphere and as carbon dioxide levels dropped, the capacity for photosynthesis (and the presence of RubisCO) spread into new groups of organisms, including terrestrial plants. Yet, despite the large changes in atmospheric composition that occurred between the first appearance of photosynthetic bacteria (several billion years ago) and land plants (500 to 700 million years ago), the RubisCO proteins found in these two groups are highly similar, both in genetic sequence and functional properties. RubisCO thus represents a molecular fossil that has functional properties that reflect the rise of photosynthesis under environmental conditions radically different from those of the present day.

Moving beyond genes and gene products, we can also see traces of evolutionary memory in the structure and organization of cellular organisms. Single-cell microorganisms affiliated with archaea and bacteria (collectively known as prokaryotes) have a basic cell architecture: most metabolic activities are loosely organized within a central fluid (the cytoplasm) surrounded by the cell membrane. By contrast, eukaryotes – which include all plants, animals, and fungi – have far more complicated cell structures: the biochemical machinery driving metabolism is contained in subcellular compartments known as organelles. The evolutionary transition from prokaryotic to eukaryotic cells (believed to have occurred as much as 2.5 billion years ago) was a major turning point in the diversification of life on Earth. The processes that led to this evolutionary transition have had an enduring impact on the emergence of multicellular life forms and the structure of ecological networks.

Symbiosis, defined as a long-term biological interaction between two different organisms, has been well documented across many organismal lineages. One of the more compelling and beautiful examples is that of the photosynthetic algae (zooxanthellae) that live inside the animal tissues of coral. These symbiotic algae give the coral reefs their striking colours and, more importantly, the ability to derive energy from photosynthesis. In the late 1960s, Lynn Margulis posited a theory for the origin of eukaryotic cells based on a process of endosymbiosis. She hypothesized that the structures inside the cells we now identify as organelles, including chloroplasts and mitochondria (sites of photosynthesis and respiration, respectively), arose from the incorporation of one cell type into another, resulting in a stable intracellular symbiosis. Although her ideas were initially rejected by the scientific community, supporting evidence accumulated over time. For example, eukaryotic organelles such as chloroplasts and mitochondria have their own genomes (separate from the nuclear genomes of the cells in which they are found), and they also have membranes, ribosomes, and modes of replication that are similar to the ancient bacteria from which they are believed to be descended. We now recognize that symbiosis is a fundamental organizing principle that records evolutionary memories at different levels of cellular complexity.

Memories encoded in the microcosmos represent 3.5 billion years of evolution, during which time microorganisms have developed metabolic pathways to harness energy and materials from the world around them. This process has fundamentally transformed the surface chemistry of Earth, and it has generated a deep reservoir of genomic diversity as microbial lineages diversified and came to occupy every conceivable metabolic niche. Today, there are an estimated nonillion (or 1,000,000,000,000,000,000,000,000,000,000) prokaryotic microorganisms on Earth. Their abundance eclipses the number of stars in the known universe, the number of neurons in our brains, and all of our synapses combined. Collectively, prokaryotic genomes define a metabolic network with the potential to encode over 15 decillion (or 15,000,000,000,000,000,000,000,000,000,000,000) genes. Although many of these genes may encode redundant information (that is, proteins with the same function), their widespread distribution within microbial lineages guards against the loss of encoded functions and provides a molecular signature of biological diversity on Earth.

Over the past several decades, unprecedented technological innovation has enabled us to read the collective evolutionary memory of the microcosmos within single cells and whole microbial communities. Scientists and engineers are beginning to see the practical benefits of harnessing these memories in the development of next-generation biotechnologies. Through such efforts, we have come to understand how microbial communities function as groups of interacting cells that have evolved modes of metabolic cooperation. This cooperation is based on modular design principles in which metabolic processes are distributed among multiple community members. In this way, a complex metabolic process (such as the conversion of organic wastes into carbon dioxide and simple nutrients) is accomplished by a consortium of cells, each capable of a particular step in the overall biochemical pathway. This division of metabolic labour among and between microorganisms interacting in complex communities is not unlike the specialization of human actors in a complex economic system, albeit a more efficient and sustainable one.

By harnessing the metabolic problem-solving power of microbial communities, we can learn to optimize our use of natural resources and the production of energy and materials. Engineered microbial communities can perform complex tasks more effectively than can single cells, and they can be more resilient in the face of environmental perturbation. For example, plant biomass provides a renewable resource for energy and materials production and has the potential to replace our reliance on petroleum products. However, the lignin and sugar polymers that make up plant biomass must first be broken down into simpler building blocks before they can be used efficiently in a biorefining context. In nature, the capacity to decompose plant biomass is distributed among multiple interacting microorganisms. Studying the metabolic networks inherent in these communities provides practical wisdom in our search for efficient biorefining processes built on the same design principles that define microbial community structure and function in the world around us.

As we enter the Anthropocene, a new epoch of human experience on Earth, the conditions for life on the planet are rapidly changing in ways that were previously unimaginable and with impacts that are difficult to predict. Shared evolutionary memories encoded and stored in the microcosmos can be traced through the information content of genes, proteins, and cellular architecture across the Tree of Life. This diversity is also expressed in the collective metabolic potential of microbial communities in natural and engineered ecosystems. By reading the genomic sequences encoded in microbial communities, we can better understand the tendency of interacting cells to evolve modes of cooperation. Memory in the microcosmos can help us reimagine innovation and redirect our own evolutionary trajectory through the development of bioprocesses that are more efficient and sustainable than any mode of energy or materials production in existence today.

# TIME, ORAL TRADITION, AND TECHNOLOGY

*Andrew Martindale, Sara Shneiderman, and Mark Turin*

HOW DO WE REMEMBER the phone numbers of friends, the birthdays of family members or, most basically, the things we're supposed to do tomorrow? Have billions of global citizens subcontracted remembering to digital devices in this technology-driven, postindustrial age? An examination of different methods of memory offers insights into how individual and collective memory works. Through a comparative lens, we see that memory is not simply passive recollection but rather a participatory form of identity construction that is highly cultural and varied. Memory emerges from the practices of individuals as well as from collectively agreed-upon recollection. Memory making occurs in an instant but records history over thousands of years.

Many human societies have relied on increasingly sophisticated recording techniques – ochre and rock, paper and pen, keyboard and screen – to document the world around them and assist them in the mission of recall. But other communities have rich oral traditions of remembering that have persisted largely unchanged over millennia. Communities that rely on oral records understand that the relationship between experience and understanding across individuals and groups defines how we know the world in particular ways. Although digital media has only been widespread for a few decades, it has considerable parallels to these systems of oral record keeping, which may explain why social media has spread so rapidly and become such an important form of knowledge making. We can see, and perhaps learn from, the ancient principles of orality as we navigate the modern world.

Within communities of shared culture, memory is constructed selectively through foundational stories; what we remember about our ancestors and ourselves is a curated subset of all possible memories. We forget the ones that do not align with our expectations. Take, for example, the collective, century-long amnesia of non-Indigenous Canadians regarding the Indian residential school system. All of us know the world through our culture, through a mosaic of experiences that functions as an ideological window into reality, resulting in a series of culturally distant bubbles, something we are witnessing in modern political debates. Although the human propensity for creating memory is shaped by technology, it is not determined by it. Record keeping, whether by rote or device, is simply the medium by which we share and recollect the historical events that enliven the stories we believe about ourselves. Traditional story forms thrive in the new digital world, and new digital practices create their own cultural communities in ways familiar to older patterns of oral transmission, suggesting that humans return to core cultural forms of memory making in any technological context.

As anthropologists, we have each had the privilege of learning from Indigenous communities with rich storytelling cultures and oral traditions, engaging with them as they have engaged with the textual and, then, the digital world. From our work with the Thangmi people of Nepal and northeastern India and the Tsimshian people of the North American Northwest Coast, we have come to understand some important aspects of communal memory. Memory is not a singular enterprise; it emerges as conscious recollection from embodied knowledge or experience, a process that often, for these Indigenous societies, links orality and performance in a powerful dynamic.

For the Thangmi community, oral traditions serve as their chief technology of memory. Members of the Thangmi community, with whom we (Sara Shneiderman and Mark Turin) collaborated for many years, were surprised by our limited power of recall. How could these outsiders, who appeared otherwise quite capable, be so bad at remembering? Why did they have to write everything down? Had they forgotten how to remember? In their own and other communities, the Thangmi have a rich reservoir of nontextual techniques embedded within oral traditions and given shape through ritual practice and performance. Indeed, Thangmi worldviews are so shaped by their commitment to oral tradition that they classify other peoples around them into two groups: those with textual religious traditions and those without.

Despite a strong cultural grounding in oral traditions, or perhaps specifically because of it, the Thangmi, like many speakers of endangered languages, have embraced new digital media with excitement. Thangmi participants in Nepal have recorded shamanic recitations of oral texts and shared them via video on social media, where they are consumed by members of the diaspora community in India, the Gulf States, and beyond. Migrant communities working in New York have recorded wedding ceremonies and songs on smartphones and then uploaded them to YouTube, where they are watched by relatives in remote Himalayan villages via cellular networks. Facilitated by digital technology, the circulation of oral tradition helps the "traditional" past converge with the practices of the present, producing transnational cultural forms of memory that rely on hybrid technological repertoires.

These insights are self-evident to communities such as the Thangmi who have circumvented the well-trodden, singular path from orality to literacy and have instead leveraged their orality directly into the digital world. Social media helps Thangmi community members rediscover and redefine their culture, bringing images, audio, and video together in ways that mirror traditional memory-making practices through storytelling. Digital communication is a new medium, but the passed-down principles of culture – the stories of who the Thangmi are – are just as meaningful because they capture how the experience of being Thangmi is produced through lived practices and performances. The agility with which members of the Thangmi community harness digital media allows for the continuity of cultural ways of knowing in the face of (indeed, with the help of) massive technological change. It also reflects a deep compatibility between traditional oral memory and the digital world of recording.

The power of memory extends beyond just the experiences of life that can be recorded and shared through various digital platforms. Humans figured out long ago that if memory-making experiences create culturally contingent ways of knowing the world, then experience can be used to craft and even discipline our understanding of the world. Cultural communities share a tacit understanding of what things mean, an understanding that emerges from sharing lifetimes of experience and memory. How we see the world is influenced by our memories of it. Where storytelling conveys the conscious understanding of cultural knowledge, performance captures the experiences that transmit the less conscious frameworks of knowledge upon which conscious understanding is built.

To illustrate the importance of cultural context in shaping collective notions of memory, consider the history of European contact with the Tsimshian people, whose waters and lands occupy the northern coast of British Columbia, Canada. The first Europeans to make contact with the Tsimshian people came armed with almost three hundred years of experience as colonizers and had strong preconceptions about the Indigenous peoples they encountered. In the European worldview, the Tsimshian, and all Indigenous peoples of North America, fit into a grand racial taxonomy somewhere between Africans, "Asiatics," and "Wild Men"; the nuanced differences among them were less important than their distance from Europeans. For their part, Tsimshian had contrasting views on the provenance of their uninvited colonizers, as is recorded in their oral records (*adawx*). The notion that Europeans, with their deathly pallor, were deceased ancestors returned to life was rejected when the newcomers did not behave with the decorum of the honoured dead. Others thought they might be porcupines because of their overall hairiness and ill temper, but many settled on the idea that they were frogs simply because there were so many of them.

The colonial encounter between Europeans and Tsimshian involved the clash of two distinct cultural systems, the head-on collision between different ways of knowing the world. While Europeans believed that humans existed as one of several biological divisions, Tsimshian classified people, along with all other natural things, by the nature of their souls. The relative correctness of these classificatory systems is less important than the resolute conviction of their adherents, a faith that emerges from the embodied cultural memory of a world both inherited and lived in. Just like the Thangmi, the Tsimshian and the Europeans knew their worlds because they experienced them as performed in ritual and everyday life.

Communities with oral traditions demonstrate our human capacity to formalize and solidify memory over generations.[1] In a world of textual documents, we might not think of oral knowledge as easily standardized, but the Tsimshian have built an intellectual edifice of orally transmitted narratives that recount the detailed history of their families since the Pleistocene. Although Western scholars and courts of law have questioned how stories told only as oral narratives can possibly be historically accurate, recent archaeological work has shown that Tsimshian narratives correctly recount millennia of history. Indeed, Tsimshian oral stories create a remembered and certain history that integrates thousands of people into a singular network of knowledge. The narratives tell the histories of genealogical lineages, migrations, and the many political and social events of their history, all framed in oral accounts of the interactions between human and nonhuman souls. Each family owns its stories and its versions of regional events, integrating them into the whole, rather like a complex weaving, to create the broad tapestry of Tsimshian history.

Memory is also key to understanding legal claims in Tsimshian society because what has been remembered is both tradition and law. Under the Tsimshian system, memory becomes both standardized and distributed. In this way, the iconic art of the Indigenous Northwest Coast serves as a mnemonic inventory of history, with each image referencing a story to be remembered. In ceremonial feasts, the narratives of history are remembered by re-enactment, creating the legal and moral precedents on which Tsimshian society is built. These texts are in some cases now written down, taking on new lives as artifacts replacing and sometimes resisted by the continuation of memory via performance.

This is increasingly the case for the Thangmi, who have sought to codify their often disparate oral traditions into singular forms, such as dictionaries and written religious texts, that can help advance their claims of recognition vis-à-vis both the Indian and Nepali states. This is a conscious act: most Thangmi agree that core cultural knowledge cannot be fully embodied in text, yet they know they must engage with textual forms to interact with the nation-states in which they live and the broader world around them. These are not contradictory "beliefs" but rather sophisticated strategies that embody the double consciousness necessary for Indigenous survival today.[2] Indeed, for the Tsimshian, the conversion of oral tradition via performance to text and recording is part of an effort to overcome the cultural genocide of the Indian residential school system, which purposely removed children from their families to prevent them from making the memories that would be the foundation of experiencing and knowing their culture.

Walter J. Ong, an American Jesuit priest, philosopher, and professor of English literature, has said that "thinking of oral tradition ... as 'oral literature' is rather like thinking of horses as automobiles without wheels."[3] This statement underscores the central question of how best to understand memory in relation to orality and textuality, challenging the presumed linear movement between these modes of recording the past. Yet the late John Miles Foley, an American professor of folklore and oral tradition, has argued that our oldest and newest technologies of communication can be considered as fundamentally equivalent in many ways. Where textual literature tends to constrain communication to a series of relationships between an author and a reader, oral and digital platforms promote iterative and lateral connections in which information is transmitted quickly within and between groups, mimicking the practice of collective rituals. In this way, Internet technologies and oral traditions share a core dynamic, enabling disparate individuals to navigate rich social networks to create patterns of meaning.[4] Both technologies foster co-creative, participatory, and ever-emergent experiences in which deeply embedded memories and new experiences are brought together to create cultural coherence.

In the rapid emergence of digital platforms, ideas and shared beliefs are largely informal. Like minds find one another along the paths of least resistance, and we have seen recent examples of how this dynamic can be exploited to target individuals with fabricated information in an attempt to persuade them towards specific political views. Our vulnerability to persuasion is less about facts and more about whether specific ideas conform to pre-existing expectations that emerge from conscious and embedded memory. As illustrated by societies that rely on oral traditions, memory, when effectively harnessed, can chart the course of social change.

The Thangmi categorically identify themselves as a people without text. They are not to be confused, however, for a people without history – or memory. Rather, it is the finely tuned practice of oral remembering that has enabled the Thangmi – like the Tsimshian – to maintain distinctive identities against all odds. For both groups, memory is maintained through cultural practices that create a sense of belonging and participation, regardless of the media in which such experiences are encoded and transmitted. Whereas the Tsimshian harness the power of memory into formal oral structures that scaffold their legal, social, and scientific scholarship across generations, the Thangmi perpetuate memory in response to the modern state, reinvigorating their oral traditions through adaptation. Both systems are changing under pressure from new technologies that reform, fracture, and reassemble the collective work of remembering via experiences of what it means to be member of a specific culture. Social media, which has transformed the world in less than a generation, uses sophisticated technology to replicate techniques that humans have employed for thousands of years to express and shape their shared cultural memory of the world around them. As they have since the beginning of humanity, the ebbs and flows of experience and storytelling will consolidate, endure, and transform who we remember ourselves to be.

NOTES

1. Andrew Martindale, Susan Marsden, Katherine Patton, Angela Ruggles, Bryn Letham, Kisha Supernant, David Archer, Duncan McLaren, and Kenneth M. Ames, "The Role of Small Villages in Northern Tsimshian Territory from Oral and Archaeological Records," *Journal of Social Archaeology* 17, 3 (2017): 285–325.
2. See Sara Shneiderman, *Rituals of Ethnicity: Thangmi Identities between Nepal and India* (Philadelphia: University of Pennsylvania Press, 2015).
3. Walter J. Ong, *Orality and Literacy: The Technologizing of the Word* (London: Methuen and Co., 1982), 12.
4. John Miles Foley, *Oral Tradition and the Internet: Pathways of the Mind* (Urbana: University of Illinois Press, 2017).

# GLOBAL 1918

*Tara Mayer and Pheroze Unwalla*

On November 11, 1918, the British prime minister, Lloyd George, stood before Parliament to announce the end of the First World War: "At eleven o'clock this morning came to an end the cruellest and most terrible War that has ever scourged mankind. I hope we may say that thus, this fateful morning, came to an end all wars."[1] Ever since the armistice was first declared, it has been marked by annual acts of remembrance across many countries of the world. These solemn events, however, tend to tell only a highly selective part of the story of 1918. While hostilities on European soil ceased that year, the end of the Great War launched a frantic reordering and repressive re-entrenchment of imperial rule across many parts of Africa, Latin America, Asia, and the Middle East. As a result, in much of the non-Western world, 1918 is not remembered as an end to war but as a period of heightened violence and oppression. Framing Remembrance Day solely in terms of Western sacrifice, endurance, and national identity obscures the global contexts that framed the causes, course, and consequences of the First World War.

The scramble for empire was a key source of conflict between the Great Powers leading up to the outbreak of war in 1914, and preserving colonial rule remained a central concern to its victors after the hostilities ended. Winston Churchill, who had served as Britain's minister of munitions during the final year of the war, noted with relief that Britain had emerged from the conflict with "its vast possessions intact" and its "Empire united."[2] Of the many theatres of intensified colonial oppression following the end of the Great War, two are especially instructive for understanding the divergent ways in which 1918 is remembered across the Global South. One is the Middle East, where the end of the First World War marked the defeat of the Ottoman Empire, the most powerful Muslim state of the time. The other is South Asia, where the British betrayal of wartime promises led to a radicalization of the nationalist movement. Together, these two examples illustrate the global dimension to 1918 and the different meanings that year holds in the memory of non-Western societies.

The Ottoman Empire had long competed with European political and economic interests in the Middle East. By the early twentieth century, however, Ottoman policy makers came to see the continued survival of the empire as dependent on the creation of an alliance with one of the Great Powers. Repeatedly rejected by Great Britain and France in the run up to 1914, the empire's Young Turk leaders allied with Germany. As the war began, the Ottoman sultan utilized his role as caliph (the symbolic leader of the global Muslim community) to proclaim a jihad against the British, French, and Russians. His call to jihad ultimately failed to make much of an impact on the Ottoman war effort. It nonetheless unnerved the British, who misjudged Islam's potential to influence the war and imperil Britain's position in India. Combined with a long-standing desire among British elites to acquire territories in the Middle East, these fears resulted in multiple conflicting British promises on the dispensation of Ottoman territories after the war.

One of these agreements, known as the Husayn-McMahon Correspondence, involved a series of letters exchanged between Sir Henry McMahon, Britain's high commissioner in Egypt, and Sharif Husayn, a descendant of the Prophet Muhammad and the Ottoman emir of Mecca, Islam's holiest city. The British asked Husayn to lead a one-hundred-thousand-strong Arab revolt against the Ottomans in exchange for an independent Arab kingdom under British protection. The number of anticipated fighters never materialized, and Husayn's revolt had little impact on the British war effort in Ottoman Syria and Palestine. The British commitment to Arab self-rule, however, was arguably contradicted by two other wartime pledges: the 1916 Sykes-Picot Agreement, which carved up the Middle East into British and French spheres of control, and the 1917 Balfour Declaration, which promised British support for the creation of a "Jewish national home" in Palestine. While there is disagreement over whether these three agreements technically conflicted with one another, the spirit of Husayn-McMahon was violated in the eyes of the Arab nationalists, who viewed (and, in many cases, still view) subsequent British actions as a betrayal.

As Ottoman defeat became inevitable, the British and French accelerated plans to tear the empire apart. With the consent of the

League of Nations, both powers established so-called mandates over particular areas, with the responsibility of establishing "a sacred trust of civilization" to aid "the natives" in eventually gaining autonomy.[3] According to their preference, the French received control of the former Ottoman province of Syria, the District of Mount Lebanon, and the Province of Beirut, openly acknowledging that they regarded these territories as permanent possessions. To consolidate their rule, the French liberally redrew borders, undermined local governance, brutally suppressed revolts, and cultivated sectarian tensions. These policies and practices are widely seen as a direct cause of the Lebanese Civil War (1975–90) and as a factor in the ongoing Syrian Civil War (2011–).

Britain's mandates included Palestine and Iraq, the latter being created from the imprudent union of three distinct Ottoman provinces. As with the French, British rule was based on weakening local institutions, brutalizing locals, and aggravating sectarian tensions that continue to plague Iraq into the present day. As regards Palestine, Britain cut the territory off from Syria, placed the Balfour Declaration at the heart of the mandate, and created the conditions for the ongoing Israeli-Palestinian conflict.

In sum, the legacies of the war are keenly felt across the present-day Middle East. The ways in which the war is remembered are, of course, far from uniform: the histories of the Armenian and Assyrian Genocides and the Kurdish partition, for example, complicate any simple narrative of Western aggression and Ottoman, Turkish, Kurdish, Arab, or Muslim victimhood. Indeed, these historical events and their politics of memory remain highly contested among scholars, politicians, and members of the general public in the Middle East and beyond. But the fractured and contested memory of the First World War across the Middle East is also, in part, a consequence of deliberate British and French efforts to stoke divisions and pit different national, ethnic, and religious groups against one another, both before and after 1918. In these ways, the memories and legacies of the Great War continue to inform relationships within and between Middle Eastern countries as well as with the Western world.

In India, the enduring memory of 1918 is, likewise, one of treachery and betrayal, epitomized most pointedly in the experience of Mohandas Gandhi, the iconic figure of the Indian independence movement. Gandhi was on his way to London when Britain declared war on Germany. Soon after, he offered his support for the unfolding war effort. Together with other Indian nationalists, he expressed "an earnest desire to share the responsibilities of membership of this great Empire, if we could share its privileges."[4] Since 1858, the year Britain formally instituted Crown rule over India, Indians had petitioned the British government for political reforms and self-governance. Indian politicians sought a degree of autonomy comparable to that enjoyed by the empire's white-settler Dominions: Canada, Australia, New Zealand, and South Africa. In Gandhi's calculation, assisting Britain in its hour of need would not only demonstrate Indian loyalty but also incur a debt that could be repaid after the war by paving the way towards Indian home rule. Instead, the British government enacted policies, such as the controversial Partition of Bengal, designed to perpetuate British hegemony over India's internal affairs.

This was not the first time Gandhi had turned wartime supporter for the British Empire. Fifteen years earlier, during his time in South

Africa, he had founded the Natal Indian Ambulance Corps, in which members of the local Indian community served as stretcher-bearers for British troops engaged in colonial campaigns. As he later wrote in his autobiography, "I felt that, if I demanded rights as a British citizen, it was also my duty, as such, to participate in the defence of the British Empire." In 1914, still in the belief that the path to emancipation was through collaboration, he sought to replicate these efforts by recruiting an ambulance corps of Indians in Britain. Gandhi was conscious of the apparent contradiction between his philosophy of nonviolence, on the one hand, and his contribution to the war effort, on the other – a contradiction that some of his erstwhile allies in South Africa and India pointed out, often bitterly. In the end, Gandhi himself never served in the Great War: he clashed with the high-handed British officer appointed to command his corps and later fell ill.

Despite this experience, Gandhi supported the British war effort after he returned to India in 1915, even as he engaged in efforts to gain independence. By 1917, as the war entered into its decisive final phase in Europe, the British government had offered significant concessions in return for Indian support. In August of that year, in the midst of the Battle of Passchendaele, Indians were promised a degree of political autonomy that came close to the coveted Dominion status. Soon after, Gandhi began to actively recruit soldiers for the British Indian Army. His efforts were sharply criticized by other members of the nationalist movement, who were incensed at the thought of Indians dying on the battlefields of Europe, East Africa, and the Middle East at the behest of their imperial oppressor. Notwithstanding such voices, Gandhi's efforts were part of a vast mobilization of manpower and

Indian soldiers during a gas mask drill on the Salonika Front. Photo by Ariel Varges, undated. Reproduced under licence by the Trustees of the Imperial War Museum.

resources across Indian society. By the end of the war, more than a million Indians had served as soldiers for the empire, with hundreds of thousands more used as indentured labourers, often digging trenches, as part of so-called coolie corps. At home, Indian industry had supplied untold amounts of ammunition, clothing, and equipment as well as revenues extracted in the form of military expenditures, war loans, and direct contributions to the British treasury. Through the loss of life, displacement, scarcity, and steep inflation, the effects of the Great War were deeply felt across India's society and economy.

Despite wartime promises for greater political independence, the British cracked down on the nationalist movement in India as soon as the war ended. Under the guise of combatting sedition, they suspended *habeas corpus*, allowed indefinite incarceration, and sharply curtailed freedoms of assembly and the press. Gandhi catapulted to national prominence through his vocal opposition to these postwar measures. Through a series of campaigns, Gandhi and other senior leaders of the nationalist movement, such as Swami Shraddhanand and Saifuddin Kitchlew, were able to capitalize on the disillusionment Indians of all classes felt about broken wartime promises. This disillusionment was substantiated six months after Lloyd George's proclamation of peace in the House of Commons, when a British colonel ordered his rifle company to fire into a crowd of unarmed protestors in the north Indian city of Amritsar, killing hundreds and injuring more than a thousand. By early 1919, it was abundantly clear that the possibility of home rule that had been granted to white-settler colonies was not going to be extended to Asian and African subject populations. Apparently, the lessons Europeans claimed to have learned from the "war to end all wars" were not going to be applied beyond Europe itself.

In European history, the century from 1815 to 1914 is often described as the long peace, a period when an alliance system kept European imperial powers from attacking one another. From a global perspective, however, the long peace appears as a time when the Great Powers exported violence across the globe by dividing the world into spheres of imperial expansion and by aggressively suppressing nationalist movements in their colonies. In those places, the First World War is not remembered as ending a long peace but rather as the explosive culmination of an ongoing struggle that had characterized much of the preceding century. Outside of Europe, Armistice Day did not mean an end to violence. Rather, the cessation of hostilities in Europe only served to intensify aggression and exploitation across much of the colonial world as the victorious powers sought to entrench and expand their hegemonies. When viewed from a global perspective, the violence that Lloyd George had described as a scourge on mankind continued to plague subject populations across Asia and Africa for decades to come.

NOTES

1 *Parliamentary Debates*, Commons, 5th series, vol. 110, November 11, 1918, col. 2463.

2 Winston S. Churchill, *The World Crisis: 1911–1918*, rev. ed. (London: 1932), 818.

3 As quoted in H. Goudy, "On Mandatory Government in the Law of Nations," *Journal of Comparative Legislation and International Law* 1, 3 (1919): 176.

4 *Indian Opinion*, September 16, 1914, as quoted in Ashwin Desai and Gollam Vahed, *The South African Gandhi: Stretcher-Bearer of Empire* (Redwood City, CA: Stanford University Press, 2016), 281.

# REWEAVING THE PAST

*Michelle LeBaron and Paulette Regan*

> Life is not what one lived, but what one remembers and how one remembers it in order to recount it.
>
> – Gabriel García Márquez

THE WRITER AND CULTURAL COMMENTATOR Maria Popova writes that "memory ... is not the pencil with which the outline of a life is drawn but the eraser."[1] This is true, she says, of both our personal memories and those recollections shared in community. But if memory is an eraser, what and who are erased? What and who stay visible? Memory, after all, does not erase evenly – it is unavoidably elective. In the ongoing dance between past and present, what is amplified, what is muted? Although memory allows us to put ourselves back into the past, we never really travel there; we recollect through the mists of current values, ideas, experiences, and cultures. Dominant perceptions of the present affect how we perceive and remember history, and remembered history can, in turn, either unsettle or reinforce the present.

In settler colonial societies, historical injustices are perpetuated not only by institutions, laws, and policies but also by what is remembered by individuals, families, communities, and nations. Memorializations of a national historical narrative in these societies necessarily reflect and reinforce colonial attitudes and injustices unless conscious, inclusive deconstructive efforts are made to create a counternarrative. Commemorations are powerful; they can act as fulcrum moments when alternative reconstructions of the past create opportunities for changed awareness and behaviour in the present. For settler colonial societies such as Canada and Ireland, commemorations can give substance and meaning to aspirations of reconciliation between Indigenous or local peoples and the colonizing power. If commemorations are to contribute to true and lasting reconciliation, then efforts must be made to seek out and give voice to what has been erased in past accounts. Reclaiming a full spectrum of voices, including those that reflect consciously chosen social justice values, is an important step towards enabling the telling of more nuanced and inclusive stories.

Commemorations often include theatre, dance, music, and the visual arts. At times, these forms of expression serve as powerful catalysts of change; at other times, they serve the status quo. Arts can portray thin, romanticized versions of the past, or they can give life to long-hidden, unsettling memories. Theatre, storytelling, and dance, for example, create possibilities for change by aesthetically exposing hierarchies inherent in economic, social, patriarchal, or xenophobic inequality. Arts present opportunities to expand and shift boundaries around what society considers relevant. When a play portrays someone who has survived violence standing up to an oppressor, audience members learn ways to challenge and resist. When a dancer movingly embodies ways to get out of tight physical spots or shows the gradual modulation of conflict into curiosity, those who watch learn new ways to move through conflict. When a piece of beautiful music touches listeners, priming them for connection and openheartedness, they may see long-standing relationships – even those darkened by enmity – differently. Particular stories, told evocatively through the arts with vivid detail and emotional currency, can change the way we remember our pasts. Any of these artistic forms, when portrayed with eloquence and compassion, can serve as vehicles for processing and releasing the rage, hurt, and trauma that accompany past injustices. The arts are uniquely powerful tools for social change, heightening emotion and sensation over thought and analysis as they create spaces for change and confront us with alternative narratives of memory.

Ireland and Canada have recently marked major anniversaries. Both countries have colonial pasts characterized by significant historical injustices. In 2016, Ireland marked the centennial anniversary of the Easter Rising, which catalyzed the eventual end of British occupation in 1922. This rebellion, led by a small group of republican activists, was brutally repressed by the British and led to significant loss of life over a six-day period. The leaders of the rebellion were executed, and other supporters were sent to internment camps in Britain. Although the rising did not immediately achieve its objective of creating an independent Irish republic, its significance looms large in the historical narrative of the country. Canada observed its sesquicentennial in 2017, evoking a colonial past that is not yet over. Colonization took Indigenous homelands and Indigenous children, attacked and devalued Indigenous cultures, and impoverished Indigenous peoples. On anniversaries like Canada 150 or the Irish Rising Centenary, retold stories either reinforce or unsettle exclusion and marginalization. How and what these public moments summon us to remember matters.

In the year of Canada's sesquicentennial celebrations, many Indigenous people and their supporters organized counter-commemorative interventions. Others participated in Canada 150 commemorations but on their own terms, telling the story of Canada in their own ways. The meme of the compassionate, beneficent settler in Canada was put up for renegotiation as Indigenous peoples across the country challenged founding settler myths of colonial nation building. This was done creatively, underlining the resilience and resurgence of Indigenous peoples. These acts of commemorative resistance were not one-off events: a broader decolonizing process was underway, in part as a result of the Truth and Reconciliation Commission, or TRC.

In the years leading up to Canada 150, the TRC began a national dialogue on Canada's Indian residential school system and that system's destructive intergenerational legacy. Alongside this process, from 2010 to 2015, residential school survivors, Indigenous and non-Indigenous artists, community groups, and organizations worked collaboratively in hundreds of arts-based commemoration projects across the country. The projects ranged from traditional and virtual quilts to medicine gardens, from totem pole and canoe carving to commemorative walking trails, from film and digital storytelling to community ceremonies and feasts. These commemoration projects were survivor-driven and culturally grounded, underpinned by collaborative research on the role of artistic practices in truth and reconciliation processes.

On the West Coast of Canada, the City of Vancouver participated in TRC events and declared itself a City of Reconciliation in 2014. In 2017, the city's year-long, Indigenous-led Canada 150+ project culminated in a traditional canoe gathering with a welcoming feast and a nine-day Indigenous arts festival. As journalist Nancy Macdonald reported in *Maclean's* on March 13, 2017, city staff were reluctant to participate in Canada 150 celebrations that reinforced colonial narratives: "Exalting Canada's colonial past two years after the [TRC] delivered its calls to action seemed regressive, and potentially harmful to the city's new relationship with local First Nations." Those who witnessed the protocols of the canoe gathering and shared food with Indigenous paddlers felt themselves becoming active participants in the reconciliation process. Indigenous peoples reframed reconciliation by using their own cultural and artistic practices for healing trauma, resolving conflict, and repairing damaged relationships, underlining the importance of culturally fluent reconciliation processes and the role of the arts in mediating and fostering such deep engagement.

Across the ocean in Ireland, another example of unsettling took place in 2016, just a year earlier. The fiftieth-anniversary commemoration of the Easter Rising in 1966 had been militaristic and triumphalist. Although more than half of the fatalities of that Easter week in 1916 had been civilians, their stories were largely untold, and women's memories and their contributions to the rising were scarcely mentioned. Leading up to the hundredth anniversary, times had changed: women's and civilians' voices could not be ignored. Shortly before 2016, women's accounts of a civilian massacre in Dublin during the rising were discovered in the archives. ANU, a theatre company, and CoisCeim, a dance troupe, produced a play called *These Rooms* based on these accounts. They turned what had been forgotten – unmentioned and unacknowledged during the 1966 commemoration – into something living, breathing, and tangible. Using theatre and dance, they sought to awaken and embody long dormant memories.

*These Rooms* opened with a summary of the rising:

> During fierce fighting ... towards the end of Easter Week 1916, the British Army suffered some heavy casualties in close combat fighting with a small, but determined, number of Irish Volunteers. On Friday and Saturday morning British troops, from the South Staffordshire Regiment, broke into houses along North King Street and exacted revenge on the local civilian population. By the following day fifteen civilians lay dead, the youngest aged just 16 years, in one of the worst massacres carried out by the British Army in Ireland in the twentieth century ... No public inquiry followed. An internal military inquiry concluded that such incidents "are absolutely unavoidable in such a business as this."

The production was set in an old house, where audience members were free to move from room to room during the restaging of the massacre. Women offered them tea and cookies as the sounds of their sons and husbands being restrained and shot in another part of the house drifted up the stairs. In a cramped bathroom, a woman washed blood off her body. A teenage British soldier in the stairwell entreated audience members to touch his hands, which had lost all feeling after executing men and boys downstairs.

In producing the work, ANU and CoisCeim wanted to bring strands of the past to light and into a current relationship with people in Dublin in 2016. They wanted audiences to wonder about the effects of the unaddressed trauma and the devastating murder of fifteen civilians uninvolved in the rising. Audiences were confronted with the smells, sounds, and sensations of unprovoked violence, bringing unexamined questions about 1916 into the present.

In Ireland in 2016, as in Canada in 2017, the arts provided powerful catalysts for new reckonings and new rememberings of old stories. We live in times when increasing awareness of colonial- and gender based oppression cries out for counternarratives of commemoration. Both the Irish play *These Rooms* and the Canadian canoe gathering intentionally highlighted the voices of the marginalized – forgotten civilians in the first instance and Indigenous people in the second. For those who witnessed them, new questions were posed. How does a violent past affect us in the present, as survivors, victims, perpetrators, bystanders, or family members? What responsibility do we bear as witnesses to personal and collective historical trauma? Are we merely spectators to others' grief, or does their pain demand an ethical response involving action? What do the stories told by Irish women and Canadian Indigenous peoples teach us about the rich potential of alternative ways of dealing with violence and conflict? These and other questions are important to consider as we commemorate the past.

Even with awareness of these questions, public commemorations remain highly political and contested. At their best, they invite us to reflect on the past with open minds, hearts, and spirits, deepening pathways of healing, reconciliation, and peace building. Alternatively, commemorations can entrench colonial and gender violence, intergenerational harms, and sectarian conflict, reinscribing dominant historical narratives while doing little to ameliorate social divisions and conflict. To avoid these destructive outcomes, national commemorations must be guided by inclusivity, tapping local knowledge and expertise and seeking broad consultation with diverse people.

Employing unsettling artistic practices in commemoration can foster moments of shared time, space, and solidarity, gathering previously neglected threads of historical memories into visibly rewoven networks of belonging. In these decolonizing moments, participants can confront and transform difficult memories and recall lived experiences of injustice and trauma. These shared experiences can deepen reciprocal empathy, respect, and compassion. Yet as Metis artist and scholar David Garneau reminds us, "Art moves us but does not necessarily move us to action."[2] Practising art can shift entrenched hierarchies towards decolonizing relations. Arts can open liminal, transformative spaces in which all involved can step into more complex memories, creating new relational foundations for joint action in the present.

NOTES

1. Mary Popova, "A Pioneering Scientist on Memory …," *brainpickings*, July 18, 2017.
2. David Garneau and Clement Yeh, "Apology Dice: Collaboration in Progress," in *The Land We Are: Artists and Writers Unsettle the Politics of Reconciliation*, ed. Gabrielle L'Hirondelle Hill and Sophie McCall (Winnipeg: ARP Books, 2015), 76.

# THE DIGITAL SHOEBOX

*Serge Abiteboul*

THE DIGITAL INFORMATION of our modern world – text, images, videos, and the like – can be stored and reproduced massively at almost zero cost. It can be easily dispersed in space to preserve it from both the natural world (fire and water) and the political one (tyrants and censors). So why does it seem so difficult to preserve our digital memory?

At first blush, we might blame changing digital formats. For example, videos of our children might be on VHS tapes, no longer compatible with current hardware. This problem is certainly real and exacerbated by rapid technological change; modern information storage formats have relatively short lifetimes, from a few years to perhaps a few decades, at most. Clearly, these recording formats are more ephemeral than Sumerian tablets or paper. Yet relatively simple solutions exist – for instance, we can use programs that translate and transcode between old and new formats, replicating information to guarantee the existence of complete and compatible copies.

We are not always aware of these solutions, and they require effort (sometimes significant) on our part. Nonetheless, the preservation of digital information is possible, and typically less expensive than the reproduction and storage of physical records.

So, what is the problem? It lies, fundamentally, in the deluge of data that forces us to choose what we want to preserve and what we are willing to forget. It is simply not possible to keep everything, nor do we necessarily want to. If we collect and archive everything, we risk turning into Funes the Memorious, a character from a story by Argentinian writer Jorge Luis Borges who remembers everything but understands nothing. The act of abstracting is a form of forgetting – we must forget some details to gain insight into the broader world around us. And herein lies the existential problem of digital memory – the choice of what to forget.

To illustrate this problem, let us consider photography. In the early days of this art form, when the production of a picture was expensive, people used their cameras sparingly, focusing their efforts on a small number of carefully chosen subjects. As the price of photographic reproduction decreased, however, output increased, and the world became flooded with images. This became particularly true with the advent of digital photography. Today, a person can easily take several hundred or even thousands of photos each year, mostly in a digital format that can be almost instantly shared around the globe. In 2016, for example, people posted 95 million photos and videos on Instagram each day. Viewing all of these images, whether in a personal collection or in a globally distributed platform, is becoming more and more of a challenge, one that requires significantly more time than many people are willing to spend.

We used to sort our information. Perhaps we had a shoebox where we kept our most precious pictures. The most organized among us made albums. Today, where are our pictures? Somewhere on an Instagram or Facebook account, perhaps, or on our phone, a computer, or external drive. Our digital information is typically spread across a combination of platforms and devices, any of which can be stolen, hacked, or damaged. Digital cloud storage simplifies our lives by releasing us from dependency on hardware we must manage ourselves. But we can get lost in the clouds. And a provider may decide, unbeknownst to us, not to archive our data beyond a few years. We change computers, we close accounts, time passes, and we lose entire portions of our memory.

What can we do? Perhaps we can create digital shoeboxes to store our most cherished information, our favorite pictures, movies, texts, and books. If we are judicious and careful, we can keep several copies of this virtual box and check from time to time that enough of them are functional. More simply, we can pay a service provider to guarantee the persistence of all this information. But, in both cases, we have to choose what to keep and what to forget. The task is painful, the problem too complex. Our only hope is that digital assistants will soon be available to help us preserve our personalized digital memory.

The opportunities and challenges of preserving digital memory are even more complex for society as a whole. Consider how the work of historians has been transformed by digitization. Once digitized, books, scrolls, and archives become available, anytime and anywhere, and access to our heritage becomes, at least in theory, universal. Some of the

European states have already digitized large swaths of their historical and cultural information. Europeana, the European Digital Library, launched in 2008, had more than 54 million digital objects by 2016, including text, images, and videos. With such initiatives, we can imagine that historians will soon have at their disposal all the information they need. They'll be able to move from one archive to another simply by changing windows on their computer screens. Their research will be freed from considerations of distance and cost. A teacher from Abidjan will think nothing of consulting primary sources in Florence.

Of course, digital connections cannot yet replace physical contact with objects or invaluable discussions with librarians or archivists. But digitization greatly facilitates research. Researchers can use optical character recognition, for example, to transform the digital image of a document into a text file, which they can then analyze and index. Historians can now more readily find and access documents on certain topics and compare authors' writing styles.

Digitization offers a particular form of immortality. The contents of an old parchment will not disappear completely if a library burns, and they will not be destroyed on the occasion of a move or on the whim of a tyrant. The ink will not fade over time. In this new digital age, traditional document preservation centres have been transformed. Take, for example, the Bibliothèque Nationale de France, established by François I in 1537 to serve as a legal deposit for copies of all books and official documents published in the kingdom. What happens to the concept of a physical repository of knowledge when most of the content produced is available on the internet in digital form? The material is potentially immortal, but it may also be ephemeral.

We must also consider what will happen in fifty years. Will researchers who want to study the world in which we live today have access to all of the information currently available on the World Wide Web? In 1994, the National Library of Canada became one of the pioneers of web archiving. Two years later, the Internet Archive foundation initiated a global archiving of web pages. Available at archive.org, the Wayback Machine allows us to go back in time to view the evolution of internet pages. A virtual tour through this site is both surprising and entertaining: the first incarnations of many long-established websites are deliciously simple and outdated.

But creating these internet archives requires heavy computing. Computer robots (so-called crawlers) surf the web, bringing web pages, images, videos, and other web-based materials into the archive. But the web is huge and the task daunting, even for an army of robots. To illustrate the immensity of the task, consider Twitter, a single website out of hundreds of millions. In just four years, from 2006 to 2010, the United States Library of Congress recovered 170 billion tweets, encompassing 133.2 terabytes of data – and this was before Twitter hosted pictures!

We cannot archive the entire web, so we must focus our efforts on areas of particular interest, usually the most popular or relevant sites, such as those dealing with elections. Even then, the web changes, new pages appear, and others evolve. We also miss considerable amounts of data contained in forms that are filled out manually. Out of the reach of web surfers and crawlers, this information can be accessed only with the agreement of the producer. We must accept that the version of the web we store in our evolving archive will never be complete or up to date. Given this fundamental limitation, we must ensure that our selective memory does not paint an oversimplified and biased picture of society, one that rewrites history by reinforcing the point of view of the dominant majority.

Beyond archiving classical digital information such as books, pictures, and web pages, there are other data that need archiving, including complex algorithms and computer programs. These products of human ingenuity and experience are part of our collective memory, our heritage, and the question of their preservation is essential. To meet this need, Inria launched the Software Heritage project in 2015 to collect, store, and share software codes in a universal, freely available archive. In just a few years, this archive has accumulated billions of files, representing a vast trove of human ingenuity and problem solving.

The preservation and archiving of digital data from both individuals and society is a quintessentially modern problem. Some of the difficulties are only temporary. Software and hardware solutions are increasingly available to facilitate the preservation of digital information and memory. But this still leaves open the question of what exactly we wish to keep. As archiving and storage become less and less expensive, the volume of data produced gets more and more unmanageable. Storage space is not the problem since miniaturization has drastically reduced the size of digital storage devices. The immense volume of data produced by CERN's Large Hadron Collider weighs nothing compared to the twenty-seven-kilometre concrete ring in which subatomic particles are accelerated to near lightspeed. However, data-processing centres require massive infrastructure. The declining cost of machinery and data storage has given us the illusion that we can keep everything. But our capacity to store massive quantities of information may make much of our data inaccessible in practical terms.

By entering the digital world, we have moved from a culture of relative information scarcity to one of information overload. With the continued explosion of data volume, perhaps the greatest challenge we face is the need to select what we want to forget. We can't address this problem without the support of computer algorithms. We must learn to use algorithms to become the archivists of our data world; we must strive to master a collective digital memory unprecedented in the history of humanity.

# INDIGENOUS STORYTELLING

*Jo-ann Archibald | Q'um Q'um Xiiem*

STORYTELLING PLAYS A POWERFUL ROLE in the way we create and recall memories. This is particularly true of Indigenous Elders who remember ways of learning the oral traditions from their ancestors, being on and with the land, and helping the younger generation learn from and with Indigenous traditional and lived stories. This concept of memory entails the development of a storied memory, the living of storied lives, the disruption of memory stories, and the awakening and resurgence of storied memories. From time to time, the Indigenous Trickster, Coyote, joins in the storytelling.

I, Jo-ann Archibald, Q'um Q'um Xiiem am from the Stó:lō / Soowahlie First Nation in southwest British Columbia and have ancestry from St'at'imc/X'x'alip First Nation in the Interior of the province. I grew up in Stó:lō territory and know some of its rivers and water systems. The rivers give the territory its name. Q'um Q'um Xiiem is my Indigenous name and means "strong, clear water."

My interest and use of Indigenous stories for teaching and learning purposes developed during my educational career, first as a school teacher and then as a university professor. I have often wondered how traditional storytellers could remember a large number of oral stories without the use of textual or digital tools. As a young school teacher, I could not remember stories without relying on written, literate forms, and I was only comfortable reading stories from books to students. It wasn't until I became a PhD student that I was able to dedicate space and time to my storied interest and to the development of my storied memory. I was drawn to Indigenous Elders and storytellers who told stories from memory. There were a few stories that I became attached to and could remember.

I want to recount one of these stories. As a form of protocol and ethics, I acknowledge the source of the Indigenous stories that I tell. Dr. Eber Hampton of the Chickasaw Nation told a trickster story at a research conference. He eventually gave me permission to use this story and to adapt it to suit my cultural context. I renamed the trickster character Old Man Coyote, because Coyote transcends time and place and has become my trickster of learning. Now it is Old Man Coyote's turn to voice his thoughts.

---

*Old Man Coyote: Searching for the Bone Needle*

Old Man Coyote had just finished a long, hard day of hunting. He decided to set up his camp for the night by starting a fire for his meal. After supper, he sat by the cozy, warm fire and rubbed his tired feet from the long day's walk. He took his favourite moccasins out of his bag and noticed that there was a hole in the toe of one of them. He looked for his special bone needle to mend the moccasin but couldn't feel it in the bag.

Old Man Coyote started to crawl on his hands and knees around the fire to see if he could see or feel the needle. He went around and around the fire. Just then, Owl came flying by and landed next to him. He asked Old Man Coyote what he was looking for. Old Man Coyote told Owl his problem.

Owl said that he would help his friend look for the bone needle. After he made one swoop around the area of the fire, he told Old Man Coyote that he didn't see the needle. Owl said that if it were around the fire, then he would have spotted it. He then asked Old Man Coyote where he had last used the needle. Old Man Coyote said that he had used it quite far away, over in the bushes, to mend his jacket. Then Owl asked Old Man Coyote why he kept going around and around the campfire when the needle clearly was not there. Old Man Coyote replied, "Well, it's easier to look for the needle here because the fire gives off such good light, and I can see better."[1]

---

Searching for the bone needle is like searching for the perfect answer or working towards one's goals. The bone needle could be a storied memory. In the story, Old Man Coyote knew how to use the bone needle, and he had used it previously. However, he did not care for the needle well and subsequently lost it. I identify with Old Man Coyote's comfort around the fire. Sometimes, we get used to doing things a certain way, and the "doing" gets easier with repeated use. However, going out into the dark or returning to places less known takes more effort and courage. This brings me back to the notion of memory and its place in storytelling.

I recall others who travelled on the story memory pathway and provided some direction for me. In the late 1980s, scholarship by Elder Harry Robinson and Wendy Wickwire introduced an important approach for remembering Indigenous stories from the Okanagan region of BC. Robinson recalled his friend, Josephine George, who told her grandson that to remember a story, you must "write it on your heart."[2] Embedding and remembering stories through the emotions – knowing them intimately – establishes a seamless connection to one's memory. Stó:lō Elder Roy Point remembered that grandparents often told stories to children at night and that the grandparents "had memories, miles and miles long with their stories."[3] He also mentioned that children listened to stories for two to three hours at a time. I wondered how the grandparents developed these extensive memories for stories and how children could listen to stories for this length of time. Developing a storied memory means that a storyteller has to keep using memory, repeatedly telling the stories to others. Sometimes, song and art forms are added to the oral telling.

Elder Dr. Ellen White, Kwulasulwut, of Snuneymuxw First Nations, from the Nanaimo area of Vancouver Island, BC, spoke to me about learning traditional Indigenous stories through structured repetition directed by family members. She was trained to be a storyteller from a young age. The role of repetition in developing a storied memory was critical. The traditional stories that she told were detailed and complex and needed to be accurately told and retold. She talked of learning parts of the story, especially the core of the story first.[4] The responsibility for learning traditional stories accurately and intimately is essential for those who become storytellers; it is part of an intergenerational pedagogy of learning.

Now, Old Man Coyote is impatient to get in a word. He knows that the only way to develop his storied memory is through living these stories, so that he can boast about his adventures. Indigenous Elders, however, had another motive and way of developing their storied memories.

Scholars such as Julie Cruikshank at the University of British Columbia, in collaboration with Yukon Indigenous Elders Annie Sidney, Kitty Smith, and Annie Ned, fostered a new appreciation for Indigenous stories through their cooperative research. Angela Sidney's statement "Well, I've tried to live my life right, just like a story" points to a way of appreciating the power and beauty of stories, which are frameworks for presenting life experiences.[5] I am forever grateful to the Indigenous Elders who kept their stories alive in their hearts, minds, bodies, and spirits. They lived the experiential stories that they told. These stories were certainly told from memory, from their experiences, where they could remember emotions, physical actions, and the context of the stories and their tellings.

Elder and Chief Dr. Simon Baker, Khot-La-Cha of Squamish First Nation exemplifies this reinforcement of storied memory. He told me many stories about spending time with his grandmother Mary

Capilano, on the land and water, helping her sell baskets to non-Indigenous people. If I asked a question, he would answer with a story. I learned to listen to his many stories and then search in them for what he wanted me to learn. Speaking from and through experience, Khot-La-Cha's activities and events became "alive" once more. I could visualize Simon's actions, identify the emotions of caring or concern, and appreciate the intergenerational closeness of his relationship with his grandmother. Living storied lives has become important for culture and language recovery and revitalization and for appreciating Indigenous people's resiliency and resistance to colonization, which has disrupted this storied memory through legislation, education, and policy.

The storied memories of Indigenous people have been assaulted through decades of colonization, decades during which disease, legislation against cultural and ceremonial practices, family and community separation through boarding or residential schools, abduction of Indigenous children for adoption and fostering purposes, and educational assimilation in public and postsecondary education systems have negatively impacted the intergenerational transmission of culture and memory.[6] However, as Stó:lō Elders remind us, these storied memories were not totally forgotten but were "put to sleep" for a while.[7] New stories of trauma were lived by generations of people who experienced the Indian residential schools that operated across Canada for more than a century. Some of these stories were brought into a broader public consciousness through the Truth and Reconciliation Commission of Canada. Oppression, genocide, and various forms of abuse were remembered and told through residential school survivor stories. But Indigenous Elders and storytellers kept embers of original Indigenous stories alive in their hearts, minds, bodies, and spirits, waiting for the time to spark the return of these storied memories.

Starting in the early 1970s, Indigenous cultural centres were established throughout British Columbia with the purpose of culture and language revitalization. In the Stó:lō area, the Coqualeetza Cultural Education Centre was established in Sardis, BC, at a location named Coqualeetza. Traditionally, its name meant a place where blankets were beaten with a stick to cleanse them.[8] Over the years, this was also the site of the Coqualeetza boarding or residential school, later an "Indian hospital" or tuberculosis sanatorium and then a cultural, political, and social service centre for Stó:lō people. Before the cultural centre was established, a group of Stó:lō people had met on a weekly basis for culture and language revitalization purposes in a member's home. They shared food and documented culture and language from memory because there were few written records that truly reflected Stó:lō perspectives. The Elders' group became a sponsored program of Coqualeetza: a coordinator was hired to organize weekly meetings. Elders were picked up and brought to Coqualeetza. One member also cooked a lunch for everyone.

I was introduced to the Elders through my work as an Indian education coordinator with the Chilliwack School District. I also enrolled in various Coqualeetza cultural learning activities, such as gathering cedar roots, making cedar root baskets, fish-drying camps, and Halq'eméylem language classes. The Elders were involved in a new social studies curriculum program for elementary grades called the Stó:lō Sitel.

I remember a short Stó:lō story "Mischievous Cubs," which the Elders had selected for the Stó:lō Sitel curriculum. One person began telling this story but could not remember it all. Over a few weeks, individuals would contribute a piece of the story or verify another's telling of a section. The Elders said that they needed to think about the story, which really meant that they had to remember or dig into their memory bank. Eventually, the assembled collective narratives became one coherent story. This shared work was a synergistic action – one remembrance "tickle[d]" the memory of someone else.[9]

Sometimes, an unplanned action would reawaken a storied memory. I recall one Elder, Ann Lindley, who said that she had been told lots of stories when she was a child, but she had not thought about the stories for many years, maybe because of colonization or various parental responsibilities. One day, the Elder's coordinator, Wilfred Charlie, called Ann to see if she wanted to go for a car ride with another Elder and himself. Ann agreed but was surprised to learn that Wilfred was taking her to a school, where he wanted Ann to tell the children stories. At that moment, Stó:lō stories were reawakened in her memory, and she subsequently told these stories to the school students.

Indigenous storied memory can be thought of as a bone needle: a useful tool for sewing, mending, and bringing Indigenous stories, storytellers, and listeners together for meaningful engagement. Traditional forms of learning and teaching through stories, intergenerational relationships, and experiences on and with land and nature have contributed to the development of a storied memory. In telling stories, Elders and storytellers portray lives that have been lived like a story, demonstrating their individual, collective, and cultural resilience and resistance to many forms of colonization. Maybe Old Man Coyote needed to keep warm by a good fire, to gather strength, to learn or practise ways of looking after his storied memory. Elder storytellers and friends such as Owl can help Old Man Coyote and each of us search for the bone needle by asking critical questions and showing us how to awaken or spark our oral memories. Today, the term *Indigenous pedagogy* is used to include traditional ways of learning and teaching. These ways are timeless. Indigenous storied memory is a form of Indigenous pedagogy for those who want to look for and then use the bone needle.

NOTES

1. Jo-ann Archibald, *Indigenous Storywork: Educating the Heart, Mind, Body, and Spirit* (Vancouver: UBC Press, 2008), 35–36.
2. Harry Robinson and Wendy Wickwire, *Write It on Your Heart: The Epic World of an Okanagan Storyteller* (Vancouver: Talon Books, 1990), 28.
3. Archibald, *Indigenous Storywork*, 75.
4. Ibid., 54.
5. Julie Cruikshank, with Annie Sidney, Kitty Smith, and Annie Ned, *Life Lived Like a Story: Life Stories of Three Yukon Elders* (Vancouver: UBC Press, 1990), 1.
6. Royal Commission on Aboriginal Peoples, *Report of the Royal Commission on Aboriginal Peoples*, vol. 1, *Looking Forward, Looking Back* (Ottawa: Canada Communications Group, 1996).
7. Archibald, *Indigenous Storywork*, 80.
8. David M. Schaepe, *Being Ts'elxwéyeqw: First People's Voices and History from the Chilliwack-Fraser Valley, British Columbia* (Madeira Park, BC: Harbour Publishing, 2017).
9. Archibald, *Indigenous Storywork*, 147.

# SELF, LOST AND FOUND

*Alison Phinney*

As a nurse, I have often had occasion to think about the failure of memory in old age – what it means, and why it matters. It is a complex medical problem and one that poses a seemingly intractable and existential question: How can one be a self without memory?

The notion of memory loss as a condition of old age is deeply seated in Western culture. "Second childishness and mere oblivion," as famously described by William Shakespeare, has come to be one of our greatest fears of aging. That fear is heard in the uncomfortable laughter when someone jokes about a "senior moment" and in the voices of health scientists worldwide when they put aside their professional differences to identify dementia as a top priority for research funding.

It is little wonder that our concern about memory loss has come to be so prominent. We know that the world's population is aging: more and more of us are living longer and, because of this, we are confronted with sobering statistics that remind us that the older we get, the more likely we are to suffer dementia. The numbers quickly become personal. We fear the loss of autonomy that comes with forgetting how to be in the world, of being unable to make independent decisions for ourselves, unable to meet our simplest needs. But more than that, the failure of memory seems like the ultimate existential threat. As the novelist Timothy Findley pointed out in his 1990 book, *Inside Memory: Pages from a Writer's Workbook*, if we cannot remember anything, how can we possibly know who we are?

Findley's question seems a natural one, but where does it come from? Why do we believe that the loss of memory equates to the loss of self? What underlying assumptions are at play in such an association? Looking back, we see that this idea has deep historical roots. The term *dementia* derives from Latin, meaning "out of mind," and Hippocrates was the first to claim a physiological basis for the condition. His view of the brain as an arbiter of consciousness persisted into modern times. In the seventeenth century, John Locke argued that self-awareness lies at the core of what it is to be a person. Locke began by positioning the self as an intelligent subject with the capacity to perceive and know the objective world through powers of thought, reflection, and reason. By building on the idea of self-awareness, he took the argument a step further, explaining that a person has the unique capacity to "consider itself as itself, the same thinking thing in different times and places."[1] For Locke, personhood hinged on memory.

This enduring idea has been the focus of debate by countless philosophers over the centuries and has gained considerable influence in modern Western culture. We understand ourselves as autonomous agents and take for granted that the sense of self is contingent on remembering that self as existing continuously through time. It is, thus, the so-called loss of self that is implicit in the way we think and speak of those with conditions such as Alzheimer's disease. People with dementia are thought to be unreliable witnesses to their own experiences, so much so that we question their very awareness of their own being. Dementia is seen to represent the breakdown of all that is uniquely human.

Understanding personhood in the Lockean sense of self-awareness serves as a powerful medical justification for seeking a cure for dementia; any disease that leads inexorably to the breakdown of personhood must be stopped. Alzheimer Societies around the world share a vision of "a world without dementia," and one group has even gone so far as to trademark that very phrase. The care of people with dementia is viewed as something that needs to happen "in the meantime." But focusing on care while waiting for a cure is not merely a stop-gap response. Caring also provides unique perspectives that foreground the person rather than the disease. Might these perspectives have the potential to shift how we view personhood and the significance of memory loss in old age, opening up rather than closing down possibilities for living a meaningful life?

Looking back through history, perhaps the most foundational idea in nursing is that of the body in care. Nurses touch, not to heal or cure but to bring comfort and support. In so doing, they bring a particular kind of attentiveness to the human body, not as a mere object or source of breakdown but as a source of meaning. The practice of care rests on an understanding that who we are as human beings, our past experiences and understanding of the world, resides in and through our bodies. The common metaphor of muscle memory reminds us that memory is more than what we can consciously dredge up and report. Memory is not just remembering facts and events about ourselves, it also resides in the body, which speaks through movement, gesture, and expression. Psychologists might describe this as procedural memory – in other words, what we do as opposed to what we think or say.

Nurses know this as well, albeit often in an unreflective, unconscious way that is rarely discussed. In a 2016 study, health researchers at the University of British Columbia, led by Gloria Puurveen, videotaped people with dementia at the end of life and then showed these videos to the attending nurses so they could watch themselves providing care. One experienced nurse watched as she dressed Ella, a frail elderly woman whose arms were stiff and contracted. Ella could not speak or move her body independently, and she did not respond to visual cues (her eyes were mostly closed, even when she was awake). But the nurse explained that she could tell what Ella wanted by the feeling in her arms: "I can feel it all the time that she's very stiff. Sometimes when you say 'relax, relax Ella' and then try [gestures moving a limb] gentle, then you can feel it sometimes she relax[es]. But sometimes, I feel it when she don't wanna change her clothes, you can feel it that she's doing like this right [gestures resistance]."[2] Ella was perhaps drawing on a kind of body memory, harkening back to what it felt like as a child being dressed by her mother. She knew how to hold her body in a way that either allowed someone to help or that communicated resistance to that help. It is clear that the attending nurse understood the signals and described them in terms of the woman's agency. She understood what Ella wanted.

Philosopher Paul Ricoer wrote in the 1970s about the idea of narrative identity, arguing that who we are is told through story. The example of Ella and her nurse forces us to take this idea a step further because it shows that our narrative identities are not simply held as stories in our heads, waiting to be expressed through words; they are inscribed in our bodies as a way of being in the world. Our identities can thus be expressed, even in the most subtle and nuanced of ways, at the end of life. These expressions of personhood and self-identity may only be evident through the practice of care – nurses come to know them by the feeling of holding a frail body in care. It is a distinct perspective, a particular way of knowing that understands personhood through its embodied expression.

A related perspective comes from the caring relationship itself. In the mid-twentieth century, Hildegard Peplau, an American nurse, defined her practice as an interpersonal process, a view that continues to be deeply influential in the world of nursing. It is reflected in the thinking of Tom Kitwood, a British social psychologist who worked closely with people living with dementia through the 1990s, when the notion of Alzheimer's disease as a social scourge demanding a cure was beginning to take hold. Kitwood pushed back against this view, arguing instead, from the perspective of care, that personhood is not so much a function of individual memory but rather of how one is perceived and treated by others. In his 1997 book, *Dementia Reconsidered: The Person Comes First*, he explains that personhood is best understood as "a standing or a status that is bestowed on one human being, by another in the context of relationship and social being." Such a view changes how we might think about the connection between memory and the self in dementia. Just as the example of the nurse helping Ella to dress reveals that the capacity for self-remembrance is deeply embodied, it also demonstrates how self-remembrance happens in a relational context. Care was not simply a matter of the nurse "doing for" Ella; rather, it was expressed through the nurse's deep engagement in "doing with." It was by virtue of the physical and embodied connection between the two women that Ella's experience *as a person* was recognized and respected.

A third perspective emerges from the work of Florence Nightingale, who was the first modern nurse to draw attention to the importance of healing environments in care. In her *Notes on Nursing*, first published in 1859, she observes that "symptoms or sufferings generally considered to be inevitable and incident to the disease are very often not symptoms of the disease at all, but of something quite different – of the want of fresh air, or of light, or of warmth, or of quiet, or of cleanliness, or of punctuality and care in the administration of diet." Looking to the surrounding environment as a focus for care was a profound insight at the time (especially considering that the germ theory of disease had not yet fully taken hold), and it remains a profound insight today. The social and physical environment of care has been long considered a core concept in nursing, opening a door to understanding personhood as the condition of being a person in the world. This explicit consideration of the environment becomes an important idea in the context of memory care. Rather than thinking about memory loss solely as a problem of individual pathology, we are prompted to consider how the experiences of people living with dementia are shaped by the physical and social world in which they are situated. A traditional nursing home where people spend their days sitting quietly alone with nothing to do stands in stark contrast to the experience of living in a dementia village – a whole community designed to provide opportunities for involvement, connection, and contribution. Such an approach (first developed in the Netherlands and now in British Columbia) recognizes the value of care environments that allow people living with memory loss to enact their right as social citizens living with purpose and meaning.

These perspectives from the practice of care challenge prevailing assumptions about the relationship between memory and the self. They remind us of the expressive capacities of our whole bodies and that we exist in relationship with one another in meaningful environments that shape our experience in powerful ways. These perspectives from care reveal a far richer understanding of personhood than the Lockean view of the self-aware subject might suggest. To answer Timothy Findley's question, even if we cannot remember anything, we may still know who we are through our experience as embodied selves living in the world with others.

NOTES

1 John Locke, *Essay Concerning Human Understanding* (1690), Book 2, Chap. 27.

2 Gloria Puurveen, "Constructing the Experiences of People with Advanced Dementia Who Are Nearing the End of Life" (PhD diss., University of British Columbia, 2016), 176.

# CONTRIBUTORS

SERGE ABITEBOUL is a computer scientist at Inria and the École normale supérieure, Paris.

JO-ANN ARCHIBALD | Q'UM Q'UM XIIEM is a professor emerita in the Department of Educational Studies at the University of British Columbia.

GEORGE BELLIVEAU is a professor in the Department of Language and Literacy Education at the University of British Columbia.

EDOUARD BARD is Professor of Climate and Ocean Evolution at the Collège de France.

DANIEL M. BERNSTEIN is an instructor in the Department of Psychology at Kwantlen Polytechnic University.

SABINE BITTER is an artist and associate professor in the School for the Contemporary Arts at Simon Fraser University.

LARA BOYD is a professor in the Faculty of Medicine and Centre for Brain Health at the University of British Columbia.

KALINA CHRISTOFF is a professor in the Department of Psychology and the Centre for Brain Health at the University of British Columbia.

ROLAND CLIFT is an adjunct professor in the Department of Chemical Engineering at the University of British Columbia.

DEBORAH A. CONNOLLY is a professor in the Department of Psychology at Simon Fraser University.

WADE DAVIS is a professor of anthropology at the University of British Columbia.

ANTHONY FARRELL is a professor in the Department of Zoology and in the Department of Land and Food Systems at the University of British Columbia.

ELEE KRALJII GARDINER is a Vancouver-based poet, author of *serpentine loop* and the forthcoming long poem memoir *Trauma Head*, and co-editor of *V6A: Writing from Vancouver's Downtown Eastside*.

MEGAN E. GIROUX is a PhD candidate in the Department of Psychology at Simon Fraser University.

JOHN GRACE is a professor emeritus in the Department of Chemical and Biological Engineering at the University of British Columbia.

SHERRILL GRACE is a professor emerita in the Department of English at the University of British Columbia.

JUDITH G. HALL is a professor emerita of pediatrics and medical genetics at the University of British Columbia.

STEVEN J. HALLAM is a professor in the Department of Microbiology and Immunology at the University of British Columbia.

GWYNEIRA ISAAC is curator of North American ethnology at the National Museum of Natural History, Smithsonian Institution.

MICHELLE LEBARON is a professor at the Peter A. Allard School of Law at the University of British Columbia.

NICOLA LEVELL is an associate professor in the Department of Anthropology at the University of British Columbia.

DIANA MARSH is a postdoctoral fellow in the Anthropological Archives at the National Museum of Natural History, Smithsonian Institution.

ANDREW MARTINDALE is an associate professor in the Department of Anthropology at the University of British Columbia.

RENISA MAWANI is a professor in the Department of Sociology at the University of British Columbia.

TARA MAYER is an instructor in the Department of History at the University of British Columbia.

CAITLIN MILLS is an assistant professor in the Psychology Department at the University of New Hampshire.

CYNTHIA E. MILTON is a professor in the History Department at the Université de Montréal.

LAURA OSORIO SUNNUCKS is a postdoctoral curatorial fellow at the Museum of Anthropology, University of British Columbia.

ALISON PHINNEY is a professor in the School of Nursing at the University of British Columbia.

PAULETTE REGAN is a scholar-practitioner and was the senior researcher and lead writer for the *Reconciliation* volume of the Truth and Reconciliation Commission of Canada's final report.

HARVEY RICHER is a professor in the Department of Physics and Astronomy at the University of British Columbia.

PATRICIA M. SCHULTE is a professor in the Department of Zoology and the Biodiversity Research Centre at the University of British Columbia.

SARA SHNEIDERMAN is an associate professor in the Department of Anthropology and the Institute of Asian Research at the University of British Columbia.

HANNA SMYTH is a PhD candidate in global and imperial history at the University of Oxford.

PHILIPPE TORTELL is a professor in the Department of Earth, Ocean and Atmospheric Science and the Department of Botany at the University of British Columbia and director of the Peter Wall Institute for Advanced Studies.

MARK TURIN is an associate professor in the Department of Anthropology and the Institute for Critical Indigenous Studies at the University of British Columbia.

PHEROZE UNWALLA is an instructor in the Department of History at the University of British Columbia.

SHANNON WALSH is an assistant professor in the Department of Theatre and Film at the University of British Columbia.

LAWRENCE M. WARD is a professor in the Department of Psychology at the University of British Columbia.

HILISTIS PAULINE WATERFALL is an adjunct professor at Vancouver Island University, a Heiltsuk knowledge keeper, and an elected Heiltsuk councillor.

HELMUT WEBER is an artist based in Vienna and a member of the research collective Urban Subjects.

JANET F. WERKER is a professor in the Department of Psychology at the University of British Columbia.

IAN WILLIAMS is a poet, fiction writer, and assistant professor in the Creative Writing Program at the University of British Columbia.

MARGOT YOUNG is a professor at the Peter A. Allard School of Law at the University of British Columbia.

Printed and bound in Canada by Friesens
Set in Monotype Sabon and Futura by Will Brown
Text design: Will Brown
Copy editor: Lesley Erickson